SUZUKI

MICK WALKER

Published in 1993 by Osprey Publishing
Reed Consumer Books, Michelin House
81 Fulham Road, London SW3 6RB

© Osprey 1993

ISBN 1 85532 298 6

All photographs by Don Morley or from
Mick Walker's collection.

Editor Shaun Barrington
Page design Paul Kime
Printed in Hong Kong

Front cover
*Suzuki and Sheene became synonymous with
success in the late 1970s and early 1980s.
Sheene's final racing year (1983), and the
bike is a factory-supplied RG500 four-
cylinder two-stroke backed by the Heron
Corporation*

Back cover
*The author testing an RG250 Gamma for
Motor Cycle Enthusiast magazine, summer
1984*

Half title page
*Riding the VX800 is like a blast from
the past, when the top bikes were
V-twins and wheelbases were long*

Right
*Apart from Sheene and Hennen, another
member of the famous Heron Suzuki squad
of the late seventies was Nottingham butcher
John Newbold. The machine seen here is his
distinctive all-black RG500*

For a catalogue of all books published by Osprey Automotive
please write to:

**The Marketing Department, Reed Consumer Books,
1st Floor, Michelin House, 81 Fulham Road, London SW3 6RB**

Contents

Introduction

Suzuki's origins stem from the year 1909 when Michio Suzuki, its young and aspiring founder, created the Suzuki Loom Company in Hamamatsu, Japan.

Business prospered as Suzuki supplied weaving looms to the burgeoning Japanese silk industry. In 1937 the decision was taken to diversify and it negotiated with the British Austin car concern to produce the Austin Seven under licence in Japan but the outbreak of war prevented production.

Following the hostilities, pedal cycles became Japan's main means of personal transport. Following the lead set by Soichiro Honda, a number of companies began to offer small 'clip-on' petrol engines which could be attached to the customer's own bicycle.

Suzuki followed this trend in 1952 by offering the Power Free, which provided a low-cost power-assisted bicycle and was effectively the launch pad for the Hamamatsu marque's future.

Within a year the Power Free was succeeded by the 60 cc Diamond Free, the larger engine helping it to cope with Japan's mountainous provincial roads.

In 1954, Suzuki announced their first real motorcycle, the 90 cc Colleda, which was also their first four-stroke. There followed a myriad of models which featured both four and two-stroke engines from the 50 cc Selped moped to the 250 cc Colleda twin. Besides motorcycles Suzuki also built cars from the mid 1950's, the first examples being powered by a small capacity two-stroke engine.

But, although the company was becoming a force to be reckoned with in the domestic market, they were all but unknown outside the national borders.

To set this right – and again following in Honda's footsteps – they entered a team of riders in the 1960 Isle of Man TT races. Success was not immediate, but after the East German MZ star Ernst Degner defected at the end of 1961, Suzuki quickly moved to obtain his riding skills and technical knowledge; the result was the company's first world championship, the 1962 50 cc series.

Track success helped showroom sales and Suzuki responded with a host of new models, notably the famous T20 Super Six. Sporting a 6-speed gearbox and 29 bhp the twin cylinder two-stroke was good for 95 mph and achieved rave reviews both in the American and European press, not to say

Barry Sheene commands a special place in any history of the Suzuki marque thanks to his double 500 cc Championship titles (the company's first) in 1976 and 1977

anything of a string of important production machine racing victories.

In America the T20 was transformed into a pukka racing machine known as the X6, whilst in Europe it became the TR250.

While the American and European press and bike enthusiasts mused about the feasibility of a possible larger capacity Suzuki, the engineers back at Hamamatsu worked relentlessly to create just such a machine. Announced in 1968 this was the T500, a machine which was to cement the Suzuki reputation as producers of motorcycles with virtually unburstable engines. Over the following years the T500 was to win many friends on both road and track and also lead, like the T20, to a full racing version, the TR500.

Still the four-stroke lobby jibed. Could Suzuki produce a two-stroke competitor to the Honda 750? They responded in 1971 with the water-cooled three cylinder GT750 which soon became affectionately known as the 'kettle' or 'water buffalo'. In race kitted trim this too joined the T20 and GT500 in Suzuki's racing hall of fame by powering many riders, including a certain young man by the name of Sheene, to victory all around the globe. Later it was Mr Sheene who gained Suzuki's first ever world road racing title in the all-important blue riband 500 cc division in 1975 on a factory four-cylinder RG500.

Meanwhile, the important American market was forcing Suzuki to alter course as stringent new pollution laws came into effect.

In 1977, the four-stroke GS range was launched with both 400 cc twin and 750 cc four cylinder models. The bikes were soon followed by a family of sports four-strokes ranging from 250 to 1100 cc.

Whilst endurance racing the derivatives of the GS/GSX range, Suzuki continued with its two-stroke engines for their GP racers and after much public demand, in 1986 offered a road going version, the RG500G. The power characteristics of this square-four powered rocket ship were not for the feint hearted but it proved a popular choice with the more experienced rider who could appreciate (and handle!) its racer-like performance. Similarly, after its 1983 launch the 50-plus bhp RG250 soon became the machine to beat in the hotly contested quarter-litre sports roadster class.

In 1985, endurance racing success prompted the launch of the GSXR range of four-stroke, four-cylinder, street-legal production racers. They used an innovative oil-cooling system and aluminium chassis to aid the low weight of these ultra quick sports bikes.

Even more recently Suzuki have come up with the superb RGV250, which ousted Kawasaki's KRIs in the important Super Sport 400 racer arena as the 1990s dawned. No doubt the famous Hamamatsu marque has even more exciting machinery waiting in the wings to add to the legend its 11,500 strong workforce has created since its baptism into the automotive sector during the immediate post-war period.

Suzuki gained a record ten 125 cc world motocross titles in an unbroken spell between 1975 and 1984, making them the most successful in this branch of off-road sport ever

First Steps

Michio Suzuki was born on 10 February 1887, in what was then the small village of Hamamatsu, some 130 miles south-west of Tokyo. Today the 'village' is a large thriving urban complex, and the Suzuki Motor Company is still based there. However, all this was very much in the future in 1909 when, at the age of 22, Michio Suzuki went into business building silk looms, which for almost three decades were to remain the backbone of his business empire.

In 1937, Suzuki built a prototype motorcycle engine and a complete car, the latter based on an existing British design. Unfortunately, peaceful development of this side of the company was not to continue, for Japan was turning away from civilian to military production in a massive armaments drive, which preceded the country's entry into World War 2 at Pearl Harbor on 7 December 1941. Therefore, Suzuki had little option but to change over to the production of various items of military equipment, a situation which was to last until the end of hostilities in August 1945.

Although its production facilities had been badly damaged, both by Allied air raids and also by a severe earthquake in December 1944, Suzuki's initial post-war prospects looked quite promising, as the occupying American forces had given authorisation for production to resume during September 1945. However, the company found their traditional loom manufacturing business in a highly volatile state, raw materials being virtually impossible to obtain. There was only one way to continue without having to lay off a substantial number of workers — to diversify and manufacture whatever was in short supply. This led Suzuki to construct such diverse items as electric heaters, farm implements and even springs for raising and lowering train windows!

All went well until January 1948, when union problems began to surface and, in August that year, it was necessary to issue new share capital to keep the company afloat. During October 1949, there was a further increase in share capital, and Suzuki sought co-operation from the unions to implement a reconstruction plan, but this was not forthcoming. The result led to two months of bitter company/union in-fighting, followed by a crippling six-month strike, which was not settled until May 1950.

Significantly, it was during this period of unrest that Michio Suzuki was to find the answer to his industrial problems. This came through his use of a pedal cycle to reach his favourite fishing location. It was then that he became convinced that the manufacture of motorised bicycles had a great future — from personal experience, he knew that they could beat having to pedal everywhere.

Early K series commuter lightweight from the early 1960s

Through experience with the pre-war motorcycle engine and car, the thought of designing or even making a complete machine did not hold any terrors. He also realised that if his company was to make a serious challenge, it would have to produce all the engine components 'in-house' and not rely on the unknown delivery times of outside suppliers.

After much consideration, work began on a prototype 36 cc two-stroke engine in November 1951. Like the majority of other Japanese manufacturers at the time, Suzuki's engine was intended as a clip-on unit which could be attached to a conventional pedal cycle. When it was launched in June 1952, it went on sale in cycle shops, because at that time motorcycle dealers were virtually non-existent. The engine was sold under the name Power Free, and it led to the design and manufacture of the much improved Diamond Free, a 60 cc version of the original engine, which went on sale in March 1953.

B100P (soon nicknamed the 'Bloop'),
circa 1967

The success of both engines was instrumental in a turn-around in Suzuki's fortunes and led to the company's first complete motorcycle, the 90 cc, four-stroke, single-cylinder Colleda (translated, 'this is it'), which was launched in May 1954. Like Soichiro Honda, Michio Suzuki realised the importance of publicity and, thus, a Colleda was entered in the second Mount Fujii hillclimb that year. Against an 86 strong field it amazed everyone by taking overall victory!

A month later, in June 1954, the Suzuki Motor Co Ltd was incorporated, and from then on never looked back. March 1955 saw the largest Suzuki motorcycle to date, the two-stroke 125 cc Colleda ST model. By this time, motorised bicycles had been deleted from the model line-up, changes in the Japanese road traffic act towards the end of 1954 finally sealing their fate.

During August 1955, Shunzo Suzuki (the founder's son) by then a senior figure in the organisation, visited North America. Even in those early days, Suzuki, along with the likes of Honda, could see that the USA represented a vast, almost untapped export market.

In November 1955, Suzuki took part in the first Asama Plains Race, and for the first time built a special 125 cc two-stroke racing machine which, unlike earlier efforts, owed nothing to the existing production models. Although unsuccessful, none the less it paved the way for the company's entry into a serious racing programme aimed at achieving world-wide recognition through sporting success on the race circuit.

The first Suzuki twin-cylinder model, the Colleda TT appeared in July 1956, and the continued growth in motorcycle sales led to the need for

increased production facilities. As a result a new plant was opened in January 1957, making Suzuki second only to Honda in size and efficiency. In February 1957, the company's founder, Michio Suzuki, retired from the day-to-day running of the organisation, having reached his 70th birthday.

After the company's performance, or rather lack of it in the 1955 Mount Asama races, Shunzo Suzuki had vowed not to allow the company to race again until it had a competitive machine. This explains why it was four long years before Suzuki tried once more. Masanao Simizu had become the first member of the racing department, which was set up in 1956, his task to develop a suitable 125 cc racing machine. In the spring of 1958, he was replaced by Takeharu Okano. A former aviation engineer and more recently a university professor of engineering, Okano was destined to play a vital role in Suzuki's future racing plans.

Rivals Yamaha gained an impressive series of results at Asama in 1957, resulting in both Honda and Suzuki increasing staff in their respective racing shops. The result was a definite programme to ensure success at the following years Asama races.

By early 1959, it was finally realised that it was a pointless exercise to attempt the modification of standard production designs, and a purpose-built racing machine was the only answer. The final result, after much testing, was the Colleda RB (Racing Bike). This featured a single-cylinder, piston-port two-stroke with a capacity of 124.63 cc (56 × 50.6 mm), which produced 10 bhp at 9500 rpm and drove through a four-speed gearbox. This was mounted in a neat twin-downtube frame with swinging arm rear suspension and vertical shocks. Telescopic forks took care of the front end, and the machine used aluminium for the fuel tank, wheel rims and break hubs.

After their success in the Isle of Man, where they had won the manufacturers team prize, Honda's dohc RC142 twin was favourite for the 125 cc class of the 1959 Asama races. Some 40,000 spectators attended, and there was a large entry of 319 riders – 45 of whom were professional works entries – in seven categories, ensuring a level of publicity never before seen in Japan for a motorcycle event.

The race took place in truly appalling conditions which soon turned the ash-based Asama course into a quagmire. Only one of the four works Suzukis finished, Ichino in fifth place behind a solid wall of Hondas. This could well have been the end of the team's efforts had it not been for a chance meeting shortly afterwards between Shunzo Suzuki and Soichiro Honda, where the latter asked Suzuki why he didn't race such a fast machine as the Colleda RB abroad. As he was to remark later, if it hadn't been for this off-the-cuff remark, it is highly unlikely that the Suzuki president would have authorised the company's first overseas racing visit to the Isle of Man TT in June the following year ... if this hadn't happened it is also likely that the story you are about to read would have been very, very different.

Lightweights

The early lightweights are described in the previous chapter and while these bread and butter machines rarely hit the headlines, they are none the less an important segment in any history of the Suzuki marque.

Here, we rejoin the story during the early 1960s when the company offered a whole range of inexpensive commuter-type machines including the MC, ME, M12, K10, K11 and SL models. To give some idea of the importance of these humble bikes is to record that in terms of the K10 and K11 models alone, a grand total of 520,000 examples were manufactured during their five- year production run.

A distinctive feature of these machines was the fitment, together with many other Suzukis, of a simple mechanically operated pump which metered the flow of oil from a separate oil tank to the vital parts of the engine. This replaced the previous petroil (petrol/oil) mixture, which Suzuki engineers realised was a nuisance for commuter-type customers.

To ensure that the system was responsive to engine requirements, the pump was linked to the throttle so that the delivery rate was increased as the throttle itself was opened, and it would not have to rely solely on the speeding up of the pump drive as the engine responded. Amongst other factors, the use of this more refined lubrication system greatly eased the possibility of an engine being starved of vital lubrication in the event of coasting down a lengthy incline, as could occur with the petroil system. In this latter case, if the throttle remained shut, no oil would reach the engine working components until it was opened again.

The pump system was not an entirely new innovation, at least one British manufacturer having employed the same basic principle as far back as 1932. But the Suzuki CCI (Controlled Crankshaft Injection) system was more refined, pumping oil direct to the vital engine internals via a series of external plastic pipes fitted with one-way valves. There was no question of having to rely upon a jet set into the side of the inlet passage, or in the carburettor itself, immediately behind the slide. The latter technique was, of course, not much different from the petroil system, even though it eliminated the need for messy and time consuming petrol and oil mixing. By ensuring a direct, positive feed of undiluted oil, the Suzuki approach to two-stroke lubrication was a significant step forward which had not, hitherto, been applied to a mass production standard motorcycle.

Following on from the popular K and M series machines came the B100P of 1964. Some nicknamed it the 'Bloop'; derived from the K10, but with dimensions of 52 × 56 mm and a capacity of 119 cc, it proved an instant best seller and remained in production until 1970.

The Bloop continued as the B120 in the 1970s; its 118 cc engine remained basically the same

The year 1967 brought another significant Suzuki lightweight, the A100 and one that was to run on virtually unchanged until well into the 1980s although initially it was only sold on the home, Japanese, market. The single cylinder was inclined well forward, so far in fact that the fins ran along its length and matched similar ones on each side of the head. Both head and barrel were in light alloy, the latter with an iron liner, and engine dimensions were 'square' at 50×50 mm, giving a capacity of 98 cc. Compression ratio was 6.5:1 and induction via a disc valve on the offside which helped the engine produce a healthy 9.5 bhp at 7500 rpm.

It was a successful format and was quickly followed by a smaller capacity version, the A50 which ran on until the end of the 1970s; including a moped version the AP50, and a sportster labelled the AS50.

During the mid 1970s Suzuki saw fit to re-introduce the 'Bloop' as the B120 Student. This exhibited little real change from the original and was

offered at a price comparable, at least in Europe, to the budget-priced commuter bikes from the likes of CZ and MZ which were then being imported from the Communist Bloc. The Student was largely replaced by the new GP series – available in 100 and 125 engine sizes – which appeared in the late 1970s. These little bikes bridged the gap between offering their owners a combination of 70 mph-plus performance (the Student could only manage 60 mph) excellent fuel economy and near-perfect handling. Previously riders had been forced to choose between speed, economy or handling in the ultra-lightweight sector of the market. The smaller GP model was particularly popular in Britain, where its sub-100 cc engine size attracted favourable insurance ratings.

Suzuki have also offered a series of 'step-thru' models. The first of these was the M30 of the early 1960s, followed first by the U50 of 1966-68, and from 1969 as the F50, eventually replaced in the mid 1970s by the FR70/50 series. These were very much in the mould of the Honda Cub, but with Suzuki's own two-stroke engine replacing Honda's ohv four-stroke powerplant.

There have of course been a myriad of other Suzuki lightweights, but the motorcycles described here are the most important and have sold in by far the greatest numbers.

Above
Suzuki's long-running GP100 commuter, first seen in 1978 and still available today

Right
Another success story for Suzuki, the GP100U has been in continuous production since the late 1970s. The GP100's 98 cc (50 × 50 mm) disc valve engine produces 12 bhp at 8500 rpm

Early Sportsters

The first Suzuki to be exported in any real numbers was the twin-cylinder T10 of the early 1960s.

The ancestry of the 246 cc (52 × 58 mm) piston port T10 can be traced back to 1956 and the company's first 250, the Colleda TT. This was soon followed by the TP and then, in 1960, by the Twinace models, the TA and TB.

The T10 was a much smoother, quieter and far less demanding bike to ride than the more sporting Suzuki twins which were to follow, and came at a time when the Japanese had yet to exploit the dark tuning secrets they were learning on the track, for street use.

The T10 had virtually no power band as such, only four gears and petroil mixture. It also had many unusual features, certainly when compared to the machines which followed. These included rotary gearchange; this allowed two methods of selecting neutral – from top gear, either back down the 'box through third, second and first in the normal way – or by changing up once more! This was probably a good idea for a commuter in the crowded stop-go environment of Tokyo, when one dab after a crash stop would get you back to neutral, but it proved a different proposition in Europe and the United States of America, where the system

The T10 twin of 1963 was an early attempt by Suzuki to score a success on the export market. Notable features included electric start, hydraulic rear brake and direction indicators

T20 Super Six (x6 in USA) set a new standard when it hit the streets in 1966

was phased out quickly on models imported to those markets.

The frame and swinging arm were of pressed steel, rather than the more conventional steel tubing construction and there was an extremely bulky starter/generator motor. This latter item was of the Siba-type and was keyed to the timing side of the crankshaft. Activation was via a push-button mounted on the offside handlebar to a relay within the voltage control box. Cranking over speed was slow because of the low torque available, there being no reduction gear ratio with the motor mounted directly to the crank.

Other features included a fully enclosed final drive chainguard, helical primary drive gears, iron cylinder barrels/alloy heads, chrome-plated tank panels, front mudguard and side panels, whitewall tyres, heel and toe gearchange lever and an hydraulically operated rear drum brake. The latter was an idea copied from German machines of the early 1950s.

Maximum speed was around 75 mph (Suzuki claimed 87 mph) and 17 inch tyres were fitted.

Popular belief would have it that the T10 was the progenitor of the now better remembered T20 Super Six (x6 in America). In fact nothing could be further from the truth, their model coding being the only point of duality. The T10 represented the end of one design family, the T20 the beginning of another; both machines shared practically nothing in the way of component parts.

The Super Six/X6 was the fastest Japanese quarter-litre sports bike of its generation and could only be challenged by Italy's Ducati Mach I. The 29 bhp, 247 cc (54 × 54 mm) engine was entirely new and major differences between it and the earlier T10 included porting, alloy cylinder barrels, pump lubrication, six-speeds, larger 24 mm carbs and no electric start.

As for performance road tests of the day obtained speeds of between 91 and 97 mph, depending upon conditions; but all were full of praise. Motor Cycle in their test of 23 June 1966 summed things up nicely, saying: 'It is a motorcycle that is hard to fault without appearing niggling. Performance, handling and superior detail finish are in harmony. It doesn't even belong to the two-fifty class, really. Its a Suzuki Super Six – and that's that!'

Cycle parts, while looking conventional enough today, were, as with the engine, a major departure from previous Suzuki roadster practice. For a start the frame was a full loop double cradle design, as opposed to the previous pressed steel affair. Paintwork was a glittering metallic candy red, blue or black, set off nicely by masses of polished alloy for items such as the brake hubs and other outer engine cases. The light-weight of the machine not only helped acceleration, but was reflected in the lean, low styling, with no superfluous equipment. The 200 mm (8 inch) twin-leading-shoe front brake, backed up by a similar sized single leading shoe affair at the rear was well able to pin the bike down from its 90-plus maximum speed.

This combination of speed and agility were also to play a leading role in the T20's performance in sports machine racing, including success in events such as the 1966 500 mile race at Brands Hatch and the Isle of Man TT.

For 1967 the smaller T200 version appeared. This was much the same but with smaller engine dimensions of 50 × 50 mm and a capacity of 196 cc. Compression was 7:1 and the engine produced 23 bhp at 7500 rpm. In some countries, including Britain, it was known as the Invader and another major difference was five, instead of six gears.

When the T20 was launched few observers realised that in less than two years Suzuki would introduce a similar bike, but with twice the capacity. The newcomer, a 492 cc piston-port two-stroke twin with bore and stroke dimensions of 70 × 64 mm, had a compression ratio of 6.6:1 and a maximum power output of 46 bhp at 7000 rpm. The overall appearance of the engine

A better overall machine than its larger three-cylinder stablemate, the GT550, the GT380 triple was a classic of its kind

was clearly based on the T20, but none-the-less it differed from its smaller brother in many detail points.

The bottom end of the 500 twin was almost a carbon copy of the 250, with needle bearings for both the big and small ends, three main bearings and four crankshaft seals. The alternator was situated on the nearside with the points outside and the helical geared primary drive on the offside. Unlike the quarter-litre T20, the newcomer employed five ratios and the gearbox itself was different.

Top end details included cylinders and heads in light alloy, with the former having replaceable cast iron liners, the cylinder head gaskets were made of aluminium. The method of changing the head and barrels differed from the 250 and there were additional bolts to provide extra strength. The early 500 twin had two 34 mm carburettors, which were replaced from the 1969 season onwards by smaller 32 mm units. In standard trim the 500 could top 110 mph, but later specially prepared racing versions, including the factory's own TR500, showed that its real potential was far greater.

The first 500 twins reached Britain in late 1967, having gone on sale in America earlier that year. The Stateside name, Titan, was not used in Britain because the leading customisers of the time, Read Titan of Leytonstone, had already registered the name. Hence the Titan became the Cobra, at least in the UK.

Certainly if compared on paper with the offerings of Honda (CB500) and Kawasaki (Mach 3), the 500 Suzuki twin seemed a very ordinary machine. But out on the road things were rather different as it offered its owner a rare blend of speed, reliability, ease of maintenance and economy, which its rivals couldn't match.

On the debit side the motor couldn't be described as smooth, in fact at times it vibrated quite dramatically. In addition the roadholding was none too secure, at least with the original equipment, Japanese tyres and rear shocks. However, substitution of either Avon or Dunlop rubber and Girling dampers together with the addition of a steering damper made a world of difference.

The exhaust note is best described as 'burbling'. On the overrun the engine tended to react as a large capacity two- stroke single, by lurching when the throttle was shut.

There were very few changes throughout its ten year life span, except for 1976, when it became the GT500A with a disc front brake, electronic ignition, a steering damper, larger capacity fuel tank and fewer horsepower. By the end of 1977 the 500 twin was no more, having made way for a new breed of Suzuki, the GS400/750 four-strokes, but it is one model which has certainly not been forgotten.

During the 1970s Suzuki continued to build various updates of its

successful 250 twin formula, including the T250 GT and X7 — the latter billed as the world's first 100 mph quarter-litre production roadster. There was also a larger version of the basic layout in 305 cc (60 × 54 mm) and 316 cc (61 × 54 mm) marketed as the T30S and T350 respectively from 1968 until 1973.

But the real interest during the early/mid 1970s lay in a pair of three-cylinder two-strokes, the GT380 and GT550. Though faster, smoother and better handling than the 500 twin, the 543 cc (61 × 62 mm) GT550 never quite made it, and certainly never matched the sales of the older design.

Brother or at least close relation to the similar GT380, the three-cylinder 550 was to suffer an identity crisis throughout its five year life span (1972-1977), Nobody was quite sure if it was intended as a tourer or sportster. In all but its final 'M' form, the 550 wasn't much quicker than the 500 twin — and much more expensive and thirsty to boot. The 1977 M model pumped out 53 bhp at 7500 rpm and achieved 115 mph, but fuel consumption at near full throttle averaged out at a mere 27.5mpg, almost as bad as the gas guzzling Kawasaki Mach 3.

This leaves the smaller three, the GT380. And in practice this was probably the best of all the early air-cooled two-stroke Suzukis. Bereft of electronic starting (which the 550 had as standard), but lighter by 34 kg (75 lb) and with an extra ratio to enable engine revs to be attuned more precisely to any road situation, it had acceleration and all-out speeds not far short of the 550's, and on B roads especially could run rings round the bigger bore bike. Other advantages included superior engine smoothness, greater economy and less weight to haul around when not on the move.

Introduced the same year as the 550, the 380 ran on another couple of years until 1979, it was sadly missed when the time came for it to be deleted from Suzuki's model line-up. Today it is rapidly becoming a classic in its own right.

Trail bikes

Like their other Japanese rivals, Suzuki's first efforts at building and marketing a street bike with off-road pretensions centred around the mildly kitted out 'street scramblers' of the mid/late 1960s, which in the Hamamatsu factory's case meant machines such as the T20C. This was a typical product of the day – essentially the standard T20 Super Six (x6 in the USA), but with high pipes, braced bars and knobbly tyres as the only real differences between the street scrambler and its road-going brother.

The first real Suzuki trail bike was the TS125. An all-new design, this first appeared in 1971 and marked a departure for the company, being a purpose-built on/off roader. It replaced the previous season's 120 cc Bearcat which, like other Suzuki 'street scramblers', was little more than a modified road machine. The TS125, by contrast, displayed strong family links with the successful motocross machines of 1970/71. The 123 cc (53 × 50 mm) piston port single-cylinder two-stroke engine worked on a four-port principle with the crankshaft carried on ball races and connected to the five-speed gearbox by helical primary gears. Carburation was taken care of by a 24 mm Mikuni. Maximum power was 13.5 bhp at 7000 rpm, enough to propel the 97.5 kg (215 lb) trailster to 70 mph.

Following in the footsteps of the 125, Suzuki quickly added the TS50 (49 cc) and TS100 (97 cc). Both these latter models retained five-speed transmissions, but featured rotary valve induction.

Next came the TS250. Like the 125 this too sported piston controlled induction and engine dimensions of 70 × 64 mm gave a capacity of 246 cc. Compression was 6.62:1 with the motor producing 23 bhp at 6500 rpm via a five-speed gearbox. Together with the 125, the TS250 was to enjoy a long life and prove to be one of Suzuki's best selling models for many years.

The TS family was soon joined by the 185 (183 cc) and TS400 (396 cc), but these never sold in quantity and were soon dropped.

The TS50, 100, 125 and 250 were continually updated and by the mid 1980s the two larger engines sported water cooling; currently only the 50 and 125 models are imported into Britain.

In the early days Suzuki also offered a series of all-terrain motorcycles, originally coded MT, and from 1973 known as the RV. These were offered in a variety of engine sizes: 50, 75, 90 and 125 cc, and were, without doubt, some of the wierdest looking two-wheelers ever made. But it has to be said that they were efficient little bikes, best summed up by a Suzuki brochure of the time which described them thus: 'No terrain's too sandy. Too rocky. Too snowy. Too wet. To bushey. Or too muddy – for the RV's fat, knobbly tyres bite in and go where you point the handlebar. It's also a perfect street

Water-cooled TS125X of 1985; it heralded a new era for the Suzuki two-stroke trial range

bike. In fact, its many different bikes in one – for as many different types of uses as you can think of. Go everywhere from now on ... on a Suzuki RV'.

Perhaps less startling, but far more relevant to Suzuki's future was its first four-stroke trail bike, the SP370 of 1978 – and also the company's first four-stroke single of any application. Designed in the wake of Yamaha's successful XT500, the Suzuki design closely followed the larger machines basic engineering principles in having a tall, sohc, black finished engine canted forward in a high-clearance frame.

The 369 cc (85 × 65.2 mm) engine ran on a compression ratio of 8.9:1 and produced 25 bhp at 7500 rpm. Features included primary kickstart, a 32 mm Mikuni carb, alloy wheel rims and gas-oil twin rear shocks.

In 1979 the SP370 became the SP400, essentially the only change (except for colour) was an increase in engine capacity to 396 cc – even the power output remained the same.

A year later and the first DR model – the 400 – made its bow. Compared with the SP370/400, the DR was a much more purposeful bike and far superior when used off-road. The year 1982 saw the introduction of Suzuki's smallest four-stroke trail bike, the 124 cc (57 × 48.4 mm) DR125, with six-speed gearbox but otherwise a decidedly muted performer.

By mid 1984 Suzuki had increased the trail lineage upwards, with the new DR600; its 598 cc sohc engine featuring twin plug ignition, 4-valves, dual exhaust system and an oil cooler to allow extended, high-speed cruising under the hottest, most arduous conditions. Maximum power output was 44 bhp at 6500 rpm and it weighed in at 136 kg (299 lb). It was offered in 'S' and 'R' versions, the latter a Paris-Dakar styled bike at extra cost.

'Mr Big' shouted the headlines of the motorcycle press in December 1987. They were referring to Suzuki's DR750, a bike which had been designed in Japan by Toshiyuki Endo but with input from Europe, most notably by former 125 cc motocross champion and multi-time Paris-Dakar winner, Gaston Rahier.

All this however was not to be enough and the press and public alike largely gave 'Mr Big' the big thumbs down. Although power delivery was smooth and the machine comfortable it was to prove a big flop. Quite simply power delivery was disappointing. The Japanese complained that noise emission levels were responsible for taking the edge off performance. Even though engine torque was increased by almost a fifth over the 600,

Desert racing DR750Z was to lead to the DR750/800 'Mr Big' large capacity single-cylinder two-stroke trail machines

Left
Current DR650 ...

Right
*... and DR350 are both leaders in their
respective classes*

this failed to make it a dramatic wheelie machine. First gear was tall like a
pure roadster, and there was plenty of forward weight. Thumper it was
not. On the road, the DR750 didn't get into its stride until 3500 rpm, not
helped by a carburation flat spot. Then followed a very flat torque curve
and steadily rising power until both tailed off as if hit by a brick wall at
5500 rpm.

Informed observers, myself included, soon concluded that the DR750 was
both soft on power and suspension, thereby ruling out serious rough stuff
(even though it had been developed alongside the factory's DR750Z 'Desert
Racer'). Even an increase in engine size to 799 cc for 1990 to create the
DR800 didn't entirely solve the problems which have been associated with
the design since its conception.

For 1991 the DR600 was replaced by the 640 cc DR650, but the biggest
shock centred around the DR350, also introduced that year. The 346 cc
(79 × 71.2 mm) engine proved to be a real gem with smooth, flexible power
delivery and clean carburation, enabling the DR350 to more than match the
capabilities of its main challenger, the Yamaha XT350. Handling is good
enough for serious off-road riding, making the DR350 what many claim is
the first real dual-purpose dirt bike. The machine is just as adept at cutting
through city traffic as it is tackling a slimy bog in the ISDE.

The DR650's not far behind, giving Suzuki a twosome trail challenge
which other manufacturers are finding hard to equal ... what a pity 'Mr Big'
just didn't deliver.

Racing

On a warm July afternoon in 1953 a tiny 60 cc two-stroke became the first of many factory entered Suzukis to contest a competitive event.

The quaintly named Diamond Free won its class in the Mount Fuji hill climb – a year later Suzuki won the event outright and the Hamamatsu marque had embarked on a racing programme which was to bring success in virtually every area of the sport.

The first foray into the international arena came in June 1960, when the factory despatched a team of riders to the Isle of Man TT. Three riders were entered in the 125 cc event – Mitsuo Itoh (who today is Suzuki's race boss), Michio Ichino and Toshio Matsumoto. However, due to a misunderstanding, the entry form was completed using the Colleda name, rather than Suzuki – so Suzuki never appeared in the 1960 TT programme!

Another problem was that Itoh was hospitalised after a practice fall. Suzuki responded by recruiting the British rider Ray Fay who, although not star material, at least had the advantage of course knowledge.

In the race, the Suzuki trio finished 15th (Matsumoto), 16th (Ichino) and 18th (Fay). Matsumoto just scraped into Bronze Replica time, averaging 71. 88mph – compared to race winner Ubbiali's MV Agusta at 86.10 mph; obviously, Suzuki still had a lot to learn.

Suzuki were envious of Honda's success in 1960, and for 1961 decided to take a leaf out of the East German MZ factory's book – rotary disc-valve induction. Things were helped somewhat by the fact that Suzuki and MZ shared the same hotel at the 1960 TT. Although MZ had tried to keep Suzuki's engineers at bay, the Japanese had at least been able to view the basics.

Even though Suzuki increased the tempo for 1961 – including signing up riders such as Paddy Driver, Frank Perris and Hugh Anderson – the results were nothing short of depressing, with more retirements than finishes.

The real breakthrough was to come thanks to the friendship between one of the Suzuki mechanics, 'Jimmy' Matsumiga, and the top MZ rider, Ernst Degner, which had developed as a result of a common interest in jazz music. This began in the Isle of Man at the Fernleigh Hotel which both teams shared for the second year running. Ultimately, it was to lead to Degner's defection later that year during the Swedish GP at Kristianstad.

Today, over 30 years on, there is still controversy surrounding Degner's defection and the part he played in transforming Suzuki's fortunes. Degner

Suzuki's first ever world championship was won by the RM62, a 49 cc midget racer capable of 10 bhp and 90 mph. Gearbox was an eight speeder, the year 1962

The 1962 disc valve RT125 single-cylinder racing engine; it couldn't match the success of its smaller brother

always maintained that he never took any MZ parts with him when he defected, but three Suzuki officials openly admit that he did. These parts – a piston, cylinder crankshaft and disc valve, plus some drawings – enabled the Japanese factory to provide their first, truly competitive Grand Prix machine in time for the 1962 season, four short months later. My own feeling is that the subsequent chain of events would appear to indicate that this is precisely what happened.

1962 was the first year of the 50 cc World Championship. The

50 cc TT was to be Suzuki's first ever classic victory, which was followed by three more wins in Holland, Belgium and West Germany, by that man Degner, who even though for obvious reasons he did not contest the East German round took his and Suzuki's first ever world title.

Degner's title-winning mount, coded RM62, had a capacity of 49.64 cc (40 × 39.5 mm) and produced 8 bhp at 10,500 rpm (increased to 10 bhp at

Factory rider Bertie Schneider on the
fearsome 250 Square Four, Dutch TT
1964. Rider's nicknamed the bike the
'Whispering Death'

12,000 rpm mid-season) and provided maximum speed of first 83 and later 90 mph. Gearbox was an eight-speeder.

The other 1962 Suzuki works racers were far less successful; the 125 single's only victory coming at the end of the season at the Argentine GP when Taveri had already tied up the championship for Honda, while the 250 twin (which shared the same 54 mm 'square' dimensions as the 125 single) also only gained one victory, which came at the non-championship Cadwell Park International in September.

The 125 had seven speeds, the 250 twin six. Both relied heavily on MZ technology, which Degner had brought with him, but even so they were largely unsuccessful. Suzuki's only real triumph had been in the tiddler class, where they disposed of both Honda and Kreidler to good effect.

Viewing the 1962 season in retrospect, Suzuki's machinery was, without

question, vastly superior to what had gone before and was to act as a springboard for the future.

For 1963, the 50 remained much as before, except for being given a rear-facing exhaust, an increase in carburettor size from 22 to 24 mm and a nine-speed gearbox. Now coded RM63, it offered 11 bhp at 13,000 rpm with maximum speed rising to 93 mph. This was good enough to retain the title, but with Anderson instead of Degner. Itoh's 50 cc TT win was the first ever by a Japanese rider.

For the 125 class, there was a brand new 123.7 cc (43 × 42.6 mm) rotary-valve twin. Coded RT63, this was produced with either rear or forward facing exhausts. By the end of 1963 it was giving 25 bhp at 12,000 rpm and 114 mph. The gearbox had eight speeds, and the machine tipped the scales at 94 kg (207 lb).

Mounted on the RT63, Anderson became a double world champion by winning the 125 cc title to add to his 50 cc success. With six victories, he beat a string of strong contenders including the likes of Luigi Taveri and Jim Redman. Suzuki went on to score four more 50 cc world titles (1964, 1966, 1967 and 1968) and two more 125 cc championships (1965 and 1970).

But it was in the 250 category that Suzuki really grabbed the headlines, albeit for technical innovation rather than race success. This was thanks to one of the most amazing motorcycles ever to grace a race paddock, the 'square four' RZ63.

The idea was to mount two of the latest 125 twins in a common crankcase, but use water-cooled cylinders and heads. The reason for the watercooling was to avoid overheating of the rear pair of cylinders, and the radiator and header tank were mounted ahead of the engine on the front downtubes. The engine had the same bore and stroke measurements as the 125 twin, giving a capacity of 247.32 cc. The power output of 52 bhp equalled that of anything in existence in its capacity class, providing a maximum speed of around 140 mph.

The RZ63's two crankshafts contra-rotated, the rear one running backwards, and the six-speed gearbox was chain driven from the centre of the rear crankshaft. The engine breathed through four Mikuni 24 mm carbs, each controlled by its own rotary disc valve.

Two exhaust pipes from the rear cylinders were tucked up under the seat, and the pair from the forward cylinders curved down under the engine. All four were flexibly mounted on springs, a system originated by MZ. Transistorised ignition was used and the rotor mounted low down between the engine and gearbox unit, on the nearside. Although the frame was an elongated version of the 125 twin, the RZ63's wheelbase was kept to 1371 mm (54 in).

Jack Findlay with the works 500 Suzuki he rode in the 1974 Senior TT

Above
Barry Sheene gets a push from his Japanese mechanic, Transatlantic Match Races, Easter 1978. Machine is a special 652 cc Suzuki RG four-cylinder two-stroke

Left
Another member of the 1978 single GB Grand Prix squad was the American Pat Hennen

Left

Pat Hennen before the start of the 1978 Isle of Man Senior TT in which he suffered a serious accident, thus ending a promising career

Above

Another American to ride for the British Heron Suzuki outfit was Randy Mamola; his 2nd place 500 cc world championship machine of 1981 is shown in the foreground with other team bikes

Like the smaller twin, the square four's watercooling was by thermo-syphon (no impeller). The light-alloy cylinder blocks and heads were made in right and left-hand pairs. Diagonally-opposed cylinders fired together at 180-degree intervals.

The debut of the square four came at the Japanese GP, held at Honda's test track, Suzuka. Degner, Anderson and Perris were entered to ride the new bike, team-mate Bertie Schneider sitting this one out thanks to a broken collarbone.

Both Degner and Perris made bad starts. In an attempt to make up for lost time, Degner tried a fraction too hard and the result was an horrific accident in which he was badly burnt when his mount caught fire. This led to a long spell in hospital and although he was to make a return to the team, his days as one of the world's premier riders were over.

Meanwhile the 250 square four was never to achieve the success to

Above

Suzuki mounted Roger Marshall leads ex-Suzuki champ Barry Sheene (500 cc Yamaha) in 1981

Left

Mick Grant (12) and Graeme Crosby (18) battle it out for victory in the Isle of Man TT. Crosby won

which its creators had aspired. The major problems centred around carburation and ignition, with plug fouling a constant source of bother. Another was the water cooling system which, when running at 90C (coolant temperature) and 12,500 rpm, was fine, but as soon as the rider needed to shut off for a slow bend, the temperature rose and the power dropped accordingly.

Even so, the 1964 version was considerably different in several aspects to the machines which had appeared the previous year at the Japanese GP. The frame had been altered and stiffened, and the wheelbase shortened. A more powerful front brake was fitted, while the engine and gearbox assembly were redesigned. The most noticeable change was in the cooling system. The original long radiator hoses, which had led from the base of the radiator to the rear of the cylinder block, had been eliminated and the water now circulated through passages in the block itself. Also, the carbs (now increased in size to 26 mm), were controlled directly by cables and not by a system of levers. Modifications which couldn't be seen included

new and narrower crankshafts and stronger gears to replace the 125 cc type used in the original design. These changes resulted in the power being raised by 2 bhp, to 54 bhp at 12,000 rpm.

One journalist at the time commenting about the appearance of the square four engine said, 'A sight to make even the most experienced race mechanic quake!' And it wasn't just the spanner men who found the Suzuki awesome – the riders did too. With its record of poor reliability and the combination of an extremely powerful engine and ultra narrow tyres (2.75 at the front and 3.00 at the rear). The machine was christened 'Whispering Death' by one of its pilots, Australian Jack Ahearn.

Another redesign was carried out at the end of 1964, but even this failed to improve matters and finally the 250 square four was withdrawn, never to be raced again.

However, this was not to be the last of Suzuki square fours, as a decade later the company introduced the RG500 (XR14) Grand Prix racer.

The 1974 RG500 had an engine layout which, in basic terms, followed its infamous forbear. However, there was one big difference – the time span of almost a decade had allowed technology to catch up with the concept. Even though it was to take two full seasons, the RG500 eventually made it as a top- line racer, unlike its older, smaller brother.

In place of the earlier one-piece cylinder block, the 500 cc engine had four separate cylinders and heads. In fact, the unit is best described as four 125 cc single-cylinder engines sharing a common crankcase and coupled by gears. Each cylinder had a separate crankshaft running in its own compartment in the crankcase – and each one could be stripped down and replaced without disturbing any of the other cylinders.

This was a most practical solution, especially since the engine was a racing two-stroke which was likely to pick-up on one cylinder occasionally, particularly during practice when juggling with various jet sizes.

With the old 250 cc RZ models, the mechanics had to strip the whole engine, disturbing all four cylinders. With the RG500, they simply tackled the one giving the problem.

On the outboard end of each crankshaft sat a thin, steel disc valve. This had a sector cut out of it which controlled the flow of mixture from the 36 mm Mikuni carburettor into the crankcase, opening and closing as it spun around. On the inboard side of each crankshaft was a gear, and all four mated with a single wide gear situated in the middle of the cranks. It was this gear-coupling which made the engine so easy to strip and assemble.

Suzuki works rider Randy Mamola (2) awaiting practice for the 1982 British GP at Silverstone. Kawasaki's Kork Ballington is race number 8

Above

New Zealander Graeme Crosby was a leading Suzuki star of the early 1980s

Left

Italy's Franco Vincini took the 500 cc world title in 1983 for Suzuki

Above
Mick Grant and an RG500

Left
The start of the 1983 Donington Gold Cup; 16 (Kite Heuwen) and 7 (Barry Sheene) are on RG500s

Porting of the cylinders was conventional race practice: with five transfer ports, one mating with a window post in the rear of the piston skirt.

Ignition was by an Hitachi electronic system on the nearside of the crankcase. Lubrication was taken care of by a pump, which supplied the main bearings and big-ends, and by three per cent oil in the fuel. Oil for the pump was carried in a tank under the saddle. Bore and stroke of the early RG500 was 56 × 50.5 mm, giving a capacity of 497.52 cc, but from 1976 onwards, this was changed to 54 × 54 mm and 494.69 cc. The engine delivered its maximum power of 95 bhp at 11,200 rpm and would rev to above this figure in the lower gears. Under 9200 rpm, the power dropped off sharply, needing the rider's full use of the six-speed gearbox. (With the later square dimensions, this situation was improved). Cooling was by water with a mechanical impeller to ensure a reliable circulation.

The early engines were prone to breaking primary drives, seizing cylinders and breaking gearboxes – the last gave the rider a particularly nasty experience.

With all these teething troubles, the best that any of the Suzuki trio could obtain in the 1974 500 cc World Championship series was fifth overall. This was obtained by the Australian Jack Findlay, with the up and coming British rider Barry Sheene – who had earlier raced an ex-works 125 and TR500 Suzukis with considerable success – sixth in the title chase.

For 1975, the power was upped to 100 bhp, with an extra 5 mph on top. The team, now largely funded by Suzuki GB, comprised Sheene, Tepi Lansivuori, Stan Woods and John Newbold. Lansivuori was the best placed rider in the World Series, finishing the season in fourth position overall.

And so to 1976. Armed with the much improved square-dimension engine, which was both more powerful (114 bhp at 11,000 rpm) and more reliable, this was to be very much the year of Barry Sheene.

Backed up by the likes of Pat Hennen, John Williams, Marco Luccinelli, Phil Read and Tepi Lansivuori, who were all mounted on the latest Suzukis, Sheene won five of the ten rounds to romp away with the 500 cc World Championship – a first for both rider and the company in the all important Blue Riband class of Grand Prix motorcycle racing.

In addition to Sheene's success, Suzuki riders also garnered the next five placings – Lansivuori, Hennen, Luccinelli, Newbold and Williams – ensuring Suzuki a clean sweep of the leaderboard in the 1976 500 cc World Championship Series.

This success was to be repeated in the following year when Sheene and Suzuki made it a double championship victory. By this time however, Yamaha had got into gear with Steve Baker, who finished second and Johnny Cecotto, who came fourth. Pat Hennen, finished the year in third

place in the title hunt, to mirror his 1976 position. Sadly, the likeable American was to suffer a serious accident which brought his career to a premature end in the 1978 Isle of Man TT (by then no longer a championship event), while Sheene was pipped for the title by the emerging challenge of Yamaha, headed by Kenny Roberts.

The following season, 1979, Sheene was down to third. For the 1980 season, he quit Suzuki to organise his own privately-backed Yamaha team which was to prove a disaster – firstly through a lack of competitive machinery, and secondly by a truly awful crash in practice for the British GP at Silverstone, which many thought at the time would be the end of his career. However, the doubters reckoned without the man's courage, which saw him stage a magnificent recovery and his eventual return to the Suzuki camp.

But if it was Sheene who had been the hero of the 1970s one man stood out in the late 1980s; this was the Texan Kevin Schwantz – who had been spotted by Sheene while racing in England. In first full season of GPs in 1988 he won two races! Then in 1989 Schwantz gained six Grand Prix victories, followed in 1990 by another five.

Suzuki's new half-litre GP racer, the RGV500, was a 499 cc (56 × 50.7 mm) V-four with twin contra-rotating crankshafts; crankcase reed valve induction; single ring pistons; six transfer and three exhaust ports. There was also an electronically controlled power valves and four 36 mm flat slide Mikuni carburettors. Maximum power of the 1990 bike was 155 bhp at 13,000 rpm, giving a top speed approaching 190 mph on the fastest circuits.

Schwantz and Suzuki entered the new decade determined to improve on the company's 15 world titles, but so far this has proved an elusive task. Suzuki have also rejoined the 250 class with a new V-twin, but again without the ultimate success of a world championship. But on past form this must surely only be a matter of time ...

Above
Keith Heuwen with victor's laurels

Right
Mark Salle gained several victories in 1985 and 1986 aboard this RG500 sponsored by Royal Cars of Sutton-in-Ashfield. Venue is Donington Park; wet tyres are fitted

Overleaf
Suzuki signed Rob McElnea for 1984. He rode the latest version of RG500 with a special carbon fibre frame

Right
Two of the most promising British privateers who rode production RG500s in the mid 1980s were Londoner Gary Lingham ...

Above
... and Nottinghamshire's Steve Henshaw

Above

Barry Sheene gives his father a lift at Ladwell Park on a RG500

Left

Mark Salle gained several victories in 1985 and 1986 aboard this RG500 sponsored by Royal Cars of Sutton-in-Ashfield. Venue is Donington Park; wet tyres are fitted

Air-cooled fours

With emission controls in America threatening to kill its major export market for their all-two-stroke range Suzuki was effectively forced to enter the four-stroke field. This came when the four-cylinder GS750 (and its smaller twin-cylinder brother the GS400) appeared at the end of 1976.

Perhaps more amazing was the fact that even though Suzuki were new to this type of engine in a modern guise, the GS750 proved not only the fastest machine in its class, but also the most compact and best mannered big bike to appear from Japan to that date.

It also was instrumental in convincing market leaders Honda that their CB750 was in need of an immediate update.

The Suzuki design showed the way forward, with its shorter stroke and double camshafts, smoother power delivery and higher engine revolutions. All this, combined with engineering standards of design which promised greater durability thanks to ball and roller bearings instead of plain bearings for the built-up crankshaft. At the other end of the scale the engine emitted fewer harmful gases thanks to a newly conceived re-cycling device.

A rear disc brake partnered a similar unit at the front, the frame was, by the standards of the day, light and relatively flex-free. Handling and road holding were also helped by a steering geometry which had been assisted by the early use of computer technology; whilst the damping for both suspension systems was improved over previous Japanese efforts in this department.

Prior to the GS750 Japanese heavyweights were not the best handling motorcycles around, and what Suzuki engineers had done was to effectively break the mould — that of accepting indifferent handling as an inevitable result of high power from multi-cylinder engines.

With a genuine 125 mph top speed, the four-cylinder Suzuki also set new standards of performance from a Japanese manufacturer for the 750 cc class. The 749 cc (65 × 56.4 mm) engine ran on a compression ratio of 8.7:1, whilst the four 24 mm Mikuni carburettors were equipped with lever-operated cold-start jets. Maximum power was 68 bhp at 8500 rpm, the maximum torque of 44 lb/ft being generated at 7000 rpm.

But it was out on the road that the GS750 really excelled. An enthusiastic tester for Motor Cycle praised its abilities in quite lurid fashion; 'That is nothing compared to the thrill of cranking the big Suzy through fast bends. You feel as if you are piloting a responsive fighter aircraft. Squirt it, and the Suzy goes like a jet. Aim it and you have suddenly got an accurate missile in your hands'.

The GS1000 was probably the best one-litre four of its day. It combined a bulletproof engine with reasonable handling and a good record in Production machine racing

Above

Classic Suzuki dohc four, the GS550E

Left

This superb sidecar outfit comprises a single-seat Hedingham sports chair and GS1000S with special forks, small wheels and Polaris fairing

The four exhaust pipes were routed into a single long, seamless megaphone silencer on each side, helping to create an appearance of narrowness, assisted by the equally narrow tyres (3.25 front, 4.00 rear).

The GS750 was followed in November 1977, by the GS1000 which Suzuki called 'A regal combination of smoothness, stability and speed'.

Not only had the dohc, five-speed engine been increased to 997 cc (70×64.8 mm) but the rest of the machine was beefed up too. This included larger 34 mm carbs, strengthened frame and swinging arm, a double disc front brake, cast alloy wheels, wider section tyres (3.50 front and 4.50 rear), larger capacity 19 litre fuel tank (formerly 17 litre), electronic ignition (points on GS750) and stronger clutch. Maximum speed was now in excess of 130 mph.

Besides the standard E model there was the shaft drive G model (see Chapter 7) and S. The latter was basically the standard chain driven model, but with a new red or blue and white colour scheme and more significantly an abbreviated fairing – actually little more than a large cowling around the headlamp, with an 'aerofoil' lip at the base and a small

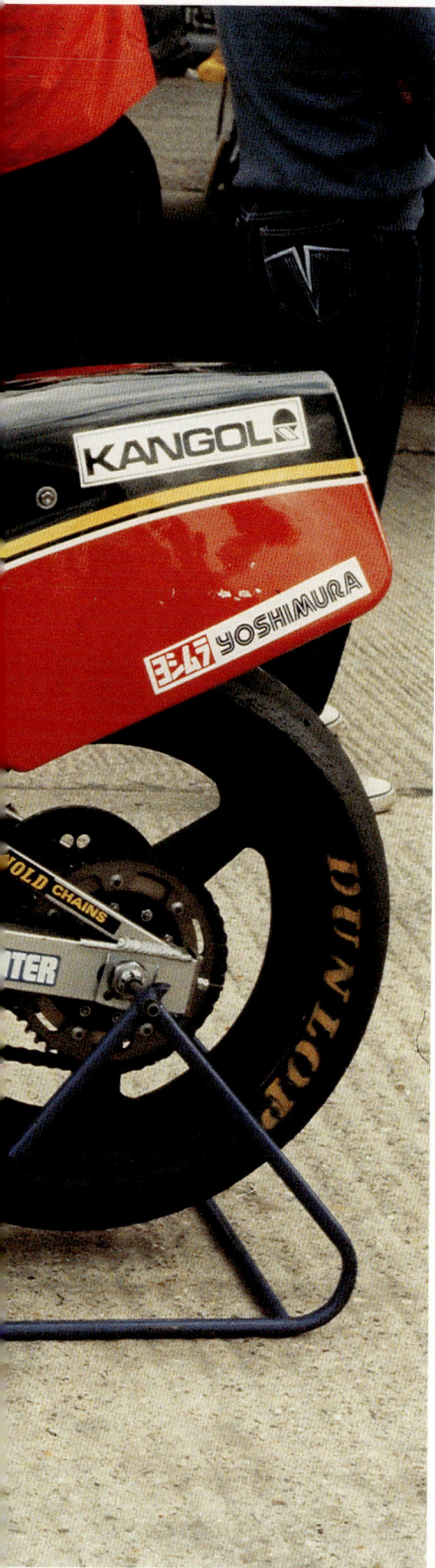

screen on the top. This fairing was too small to have any real practical use, except to keep some wind off the riders chest.

Other changes included air-adjustable front forks and rear shocks with a wide range of damping in addition to normal multi-position loadings. The variations possible, adjustment being made more difficult by the lack of an equalizer pipe between the fork legs, meant that any ride from plushy armchair comfort to racer-like tautness could be achieved.

A new instrument console, made possible by the fairing, incorporated a clock and an oil temperature gauge in addition to the fuel gauge featured on the E and G models.

Power output of the E and S were identical, at 87 bhp with the engine spinning over at its 8000 rpm redline.

The year 1981 saw the more powerful GSX1100 and 750 models. Both featured the new generation, 16-valve 'Twin Swirl Combustion Chamber' (TSCC) cylinder heads. The latter being the work of engineer Shirasagi. In the interests of the twin swirl arrangement, its designer had abandoned the conventional hemispherical chamber pattern of the 8-valve GS models, reshaping it into a square profile with various angles introduced courtesy of computer aided research, to agitate the incoming mixture for more complete combustion. A flat-top piton with four small valves helped the squish effect, while Suzuki claimed an overall improvement in efficiency of no less than 20 per cent.

Modifications to the double overhead camshafts layout included forked tappets running off the cams to actuate the additional valves, with an incidental advantage by way of screw adjustment for clearance in place of the original time-consuming system of replaceable shims.

The suspension had also been updated. Already in the vanguard of innovation in this area, through adoption of air assistance at the front, Suzuki refined their system by incorporating isolating valves for each leg of the fork to ensure that, once inflated by a common pressurizer, there would be no danger of a unilateral leak collapsing the entire fork. In addition variable pre-set damping was introduced in which any of five settings could be selected by turning a knob at the base of the fork legs. This knob caused a long internal rod to revolve, with different sized holes lining up to regulate damper-oil flow. All very complicated; and guaranteed, of course, to ensure that a GSX1100 owner spent a shade too much time on meticulous pre-ride preparation.

Instrumentation followed car, rather than bike fashion, by featuring a printed circuit layout in the speedometer/tacho console in which the outline of a motorcycle was illuminated by warning lights at salient points. A failure of electrics or fluids would cause an appropriate 'idiot' light to

John Newbold's 997 cc Heron Suzuki F1 bike

glow in response. Owners verifying the 11.5 second standing start quarter mile and 138 mph plus maximum velocity were at least likely to receive some warning if one of the superbike's life-support systems was about to malfunction!

The engine and running gear were developments of the tried and tested GS750/1000 line, but with subtle differences, most notably in the top half of the motor as already described and the machines styling.

The colour scheme could almost be described as bold, having a mainly bright red finish, with gold lettering; the square headlamp and its surround were ugly, but distinctive.

Performance equalled anything available in 1981, ensuring excellent sales; the 1075 cc (76 × 66 mm) motor making 100 bhp at 8700 rpm. A performance setter in its day, 100 bhp is regarded today as relatively tame.

There was also a GSX750. This was essentially the larger bike but with a

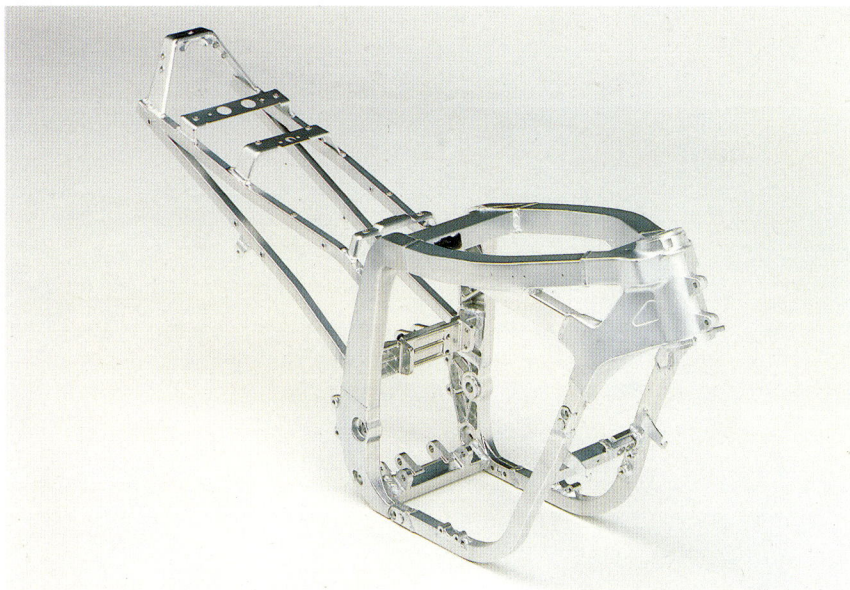

smaller, less powerful engine. The 67 × 53 mm dimensions were notably shorter stroke than the earlier GS750 and gave a capacity of 747 cc. The 1981 GSX750 lived very much in the shadow of its larger brother and sales suffered as a result. So poor in Britain was its showroom performance that the importers were forced to off-load a whole shipment at a loss making price which was below that of other makers 550s. Not that it was a poor motorcycle, just everyone who wanted to buy a GSX plumped for the 1100.

In point of fact Suzuki had its own GS550E, but unlike the GSX750 it sold for several years virtually unchanged alongside the GS750 and GS1000/1100 – outliving the larger bikes until it was finally axed in 1986.

In the autumn of 1981 Suzuki introduced a whole batch of new air-cooled four-cylinder models with futuristic styling known as the Katana, in a number of engine sizes.

The Katana's used a styling exercise which was laid out for Suzuki by the Target Design Company who had offices in West Germany and Britain, and was headed by one-time BMW designer Jan Fellstrom.

The top models were the GSX1000S Katana and the GSX1100S Katana (there was also a smaller 650 with less radical styling). Both employed a totally new set of clothes which transformed their appearance quite drastically, with a series of sweeping angles and curves, together with a sharply pointed fairing with integral square headlamp, black exhaust system and two-tone seat.

The press and public alike really fell for the big Kaftans – so much so, that, by public demand, a 250 version of the original concept is one of the

most popular bikes in Japan a decade later in the early 1990s!

The 1000 Katana used the latest version of the familiar 997 cc dohc engine with the power output upped to 108 bhp at 8500 rpm and Suzuki claimed a top speed of 140 mph.

The year 1983 saw the GSXES range comprising, 550, 750 and 1100 cc versions. The smaller mount was billed as 'Probably the most exciting bike in this years Suzuki range'. It featured an all-new alloy square section frame, half fairing, black exhaust system, 16-valve engine, anti-dive forks, "Full Floater" monoshock rear suspension and 16 inch front wheel. The 572 cc (60×65 mm) motor produced 61 bhp at 9500 rpm giving a top whack of 122 mph.

The 750, actually the best of the bunch, had much more rounded lines than the singularly angular 550. The 747 cc (67×53 mm) engine pumped out 83 bhp and had a top speed of 134 mph. Differences compared to the 550 included a square section steel tube frame and a chrome-plated exhaust system.

Finally the 1100ES featured the 16-valve dohc 1074 cc engine and was rated at 111 bhp. Maximum speed at 143 mph was ten more than the 750ES. Notably the front wheel was a 19 inch component, not 16 inch as on the two smaller ES models.

Two other air-cooled Suzuki fours are worthy of mention, but for different reasons. The first is the XN85 Turbo; introduced in a blaze of publicity this it was to prove a major flop. The idea was good – small engine, big power. But, like the Honda CX500/650 Turbo and Yamaha

XJ650, the XN85 wasn't a particularly good performer, 85 bhp/130 mph and attracted few buyers. I tested an example for Motorcycle Enthusiast in 1985 and found it good, but not good enough.

The other bike was the GSX1100E in its final guise. This was also available in EF (fairing) mode. Its outstanding feature was the engine's power – 124 bhp and a maximum torque figure of 77.4 ft/lb, which combined with a dry weight of only 232 kg (511 lb) meant serious performance, at least in a straight line. Handling and roadholding didn't stand up to that of the racer inspired GSXR series which by the time the final GSX1100E appeared were also available.

Other features included a 630 O-ring final drive chain, 16 inch front wheel (17 inch rear), 20 litre fuel tank, a massive 190 mm 12 volt 60/55 watt headlamp, black exhaust system, oil cooler, and a box section frame.

But time was fast running out for the large capacity air-cooled system four-cylinder engine concept, first in the shape of the air/air-cooled GSXR-type unit and thereafter the threat of ever increasingly stringent legislation on noise levels was to bring in water-cooling during the early 1990s – the first of which was the 1992 model GSXR750.

Above
Last of the line, the GSX1100E offered 123 bhp and rocket-like acceleration

Left
With the new for 1983 GSX750ES, the Japanese finally had a large capacity bike which, to quote a Motorcycle Enthusiast test, 'sticks to the road like glue over ripples, bumps and white lines, and steers like a dream'

Touring

The first ever, big bore Suzuki designed expressly as a touring bike was the 1979 GS850G. Developed from the 750 four with some features of the GS1000, it was also the company's first shaft driven model.

Suzuki engineers paid much attention to the design of the shaft drive system, not wishing to inherit the clunky, less-than-perfect gear change from European touring models like BMW and Moto Guzzi.

For starters they provided the GS850 with a gear-shock damper to absorb much of the 'direct drive' found in the shaft drive system and arranged a 90 degree turn in transmission for the shaft pick up. In addition, power shock was absorbed by means of a rear hub damper rubber mounted on the rear wheel housing, preventing it from being transmitted to the shaft. The result was a much smoother gear change and improved handling thanks to the virtual elimination of the dreaded torque reaction found on the European bikes. Handling and road holding were also helped by the use of tapered needle roller bearings in the swinging arm pivot.

To provide a fair level of performance it was necessary to increase the capacity to 843 cc (69 × 56.4 mm). This was achieved by giving the GS750 cylinders an increase in bore size of 4 mm, but the stroke remained

Suzuki's top-of-the-range sports/tourer, the 1127 cc GSX1100F offers blistering performance and long distance comfort

The GSF400 features a jewel-like 398 cc dohc engine with six- speed gearbox and exceptionally well made steel chassis which provides superb roadholding

unchanged. This meant an 11 bhp power increase, to 79 bhp (at 8500 rpm) – needed as dry weight had risen from 213 kg (470 lb) to 253 kg (558 lb), thanks in the main to the shaft drive layout.

The oversquare engine dimensions with hemispherical combustion chambers, large 36 mm diameter valves and four Mikuni 32 mm carbs provided the GS850 with a top speed of almost 130 mph – excellent for its day. It was also to prove a popular and long lived machine thanks in no small part to its comfortable riding stance and virtually bullet-proof engine.

The success of the 850 tourer led directly to the GS1000G for the 1980 model year. Somehow this never fully replaced the original shaft drive model, even though it had another 10 bhp. It is worth commenting that its main sales came in the USA and it did cure a major drawback of the original GS1000. This suffered, like other very powerful large capacity

multi-cylinder motorcycles of the late 1970s which employed conventional transmissions, by a final drive chain which would stretch at alarming rates. This was before the days of the 'O' ring chain; the result being that the chain would slacken, requiring frequent adjustment, and after too few miles (usually no more than 3-4000 in many cases) it would have to be replaced.

Compared with the 850, the 1000 shaft drive model was much more of a handful for its rider, thanks mainly to the extra power of the larger engine. It took a brave (read foolhardy!) rider to extract the full performance from the 1980 GS1000G. Remember that large capacity Japanese bikes didn't go round corners as they do now... The 1000G eventually became the 1100G (1983), but it was otherwise largely unchanged.

Suzuki completed a trio of shaft drive four-cylinder models with the

Featuring a detuned version of the original GSXR air/oil cooled engine, the GSX750F is an unsung hero of the Suzuki range, providing a good mixture of performance (105 bhp), comfort and safe handling

Budget priced six-hundred GSX600F is no mean performer in its own right

introduction of the GS650 (1980 USA, 1981 Europe). It was available in both conventional and Katana versions, each employing the same 673.8 cc (62 × 55.8 mm) dohc, five-speed mill. With 72 bhp at 9500 rpm on tap and a considerably lower dry weight of 220 kg (480lb), the GS650 shaft drive models were no mean performers. This meant excellent acceleration for the engine size and a maximum speed of 122 mph; however, fuel consumption was another matter, and could drop all too readily to 30-35 mpg when the bike was ridden hard.

Although Suzuki offered several middleweights such as the GN400 single and GS400/425/450 and GSX400 models during the late 1970s and early 1980s which could be labelled 'touring', the next purpose-built models didn't arrive until the late 1980s. These included the GSX110F (1988) and GSX600F (1988), GS500 (1989), GSX750F (1989), VX800 (1990), GSF400 (1991) and GS1100G (1991). Some are more touring than

others, but they are all able to travel long distances with a fair degree of comfort – certainly far more than the raw-boned super sporting GSXR range for example.

The GSX1100F was given its world premier in September 1987, being heralded, at least by Suzuki, as a machine which 'combined race performance with streetbike comfort and versatility'. Unfortunately, it somehow missed the mark, even though its engine could boast the highest torque figures up to that time for a Suzuki model, delivering a massive 82.5 lb/ft at 7000 rpm. The engine was based on the air-oil cooled GSXR1100, but with a 2 mm larger bore and a 1 mm longer stroke to provide 1127 cc. There was also a new computer-controlled ignition system and a larger volume, four-into-one exhaust system.

The most notable feature of the GSX1100F was its 'Power Shield' windscreen which could be raised or lowered at the touch of a button.

While performance and comfort were there for the taking, sadly handling and road holding were not. The GSX1100F was given 16 inch wheels front and rear and, as Moto Guzzi had found to their cost with the 850T5, this combination did not suit a large, high-speed motorcycle. Not only was handling less than perfect at high speeds, but cornering clearances suffered badly.

Since its introduction the 1100F has received a series of small but significant changes which have made it a better bike, but one which has found it difficult to compete against rivals such as the Yamaha FJ1200, Honda CBR1000, let alone the awe-inspiring Kawasaki ZZ-R1100.

Much the same can be said of the 748 cc (73 × 44.7 mm) GSX750F and 599 cc (62.6 × 48.7 mm) GSX600F, but unlike the 1100F both are competent all-round motorcycles with no real faults and have remained virtually unchanged over the years. The main sales weapons having been a low purchase price, rather than high performance or hi-tech features. (The smaller bike seems set to be phased out in favour of the new high performance RF600R).

The GS500 is also a budget priced bike, but a damn good one. The story starts back in 1989 when the GS500E superseded the GS450 (which in turn superseded the GS425, which superseded the original GS400). It wasn't so much a new bike as an old bike getting better. The engine is basically an air-cooled, dohc, parallel twin with two valves per cylinder – a bored out 450 in other words. Specifically, the GS500 displaces 497 cc from a bore and stroke of 74 × 56.6 mm, the compression ratio is 9:1 and it breathes through a pair of 33 mm Slingshot carbs. Power is transmitted to

Suzuki's VX800 has a four-valve per cylinder V-twin engine, water-cooling, 5-speed gearbox, twin shock rear suspension and shaft final drive

VX features an 805 cc (83 × 74.4 mm) V-twin engine with vibration-beating 75 degree crankpin configuration and tremendous mid-range torque

the rear wheel via a six-speed gearbox and a chain final drive.

Whilst 42 bhp is hardly class setting, it none-the-less manages to provide a fair level of performance on the GS500. A top speed of 115 bhp is possible without too much effort and its standing quarter times are good enough to see off all but the most enthusiastically-driven supercar. But the engine's most impressive feature is its smoothness, a revelation for an old fashioned parallel twin!

The motor is housed in a double cradle steel tube chassis that is supposed to ape a race-replica aluminium beam frame. Combined with efficient if not fancy 36 mm front forks, only the rear suspension lets things down. However, the GS500 is immensely flickable and a real joy to ride on twisty country lanes.

The riding position is excellent, but maybe a shade too sporting for serious touring, even though the 17 litre (3.8 gal) tank gives the GS an effective range of at least 200 miles. Home maintenance, unlike many modern machines, is relatively simple, with everything falling easily to hand. Summed up, the GS500 is cheap to buy, cheap to run and it is fun with a capital F.

Another fun bike is the jewel-like GSF400 Bandit. If the GS500 is sure footed around corners, the Bandit is even better, sticking like Super Glue.

The water-cooled, dohc 16-valve four-cylinder engine displaced 398 cc (56 × 40.4 mm) and is notable for its fine finish − superior to virtually any other Suzuki model currently in production. It also boasts 50bhp, which is enough to propel this 168 kg (370 lb) mini rocket to almost 120 mph − not bad for a mere 400 without any form of streamlining!

The GSF's only failing is its lack of engine torque; to extract any sort of performance it has to be thrashed, and serious two-up work is out. But as a one-person bike you really can have a bundle of fun, and the styling is truly beautiful.

The two final Suzuki touring bikes both suffer through poor styling; these are the VX800 and the GSX1100G.

The VX is basically a VS Intruder custom motor with higher gearing plumped into a rather boring set of clothes ... if only the factory had used the cycle parts from the GSF400.

Even so, the 805 cc VX (83 × 74.4 mm) VX is a grossly under-rated bike let down only by its low-tech chassis and suspension, and by the lack of twin discs at the front, which are really needed when two-up and carrying luggage.

Otherwise, it's a lazy, easy going sort of bike which is in direct contrast to the frantic riding style needed to get the best out of the GSF400. Its also a comfortable bike, even if the seat looks to be on the thin side. Maximum speed is around 125 mph, thanks to a small frontal area rather than lots of power.

In my opinion, Suzuki made a major error by going for the 'Modern Classic' look with its 1970s type frame and cycle parts, rather than employing 1990s state-of-the-art technology. As it is, that lovely torquey engine isn't quite matched by the rest of the bike.

The final bike is the GSX1100G. This is a new edition of an old Suzuki theme; the dohc, air-cooled four-cylinder motor with shaft drive, but unlike the VX800 the GSX at least has the advantage of a more modern frame design, with single shock 'Full Floater' rear suspension.

The 1127 cc four gives tremendous torque, the large dual seat is one of the most comfortable around, the suspension soaks up bumps like a sponge, but on the debit side it is heavy (261 kg), lacks any form of weather protection as standard equipment, the fuel tank is far too small, and the bike's styling is, well, awful; hardly surprising therefore that sales of the GSX1100G have proved disappointing to say the least since it went on sale early in 1991.

None of Suzuki's current crop of tourers can be judged sales successes in the same way as bikes like Honda's Gold Wing, Yamaha's FJ1200 or Kawasaki's GT550/750, but they do offer interesting alternatives ranging from the tiny GSF400F through to the muscle bound GSX1100G.

RE5 Wankel

Some ten years ago it seemed that every automobile manufacturer worthy of the title was offering a turbo to their customers; and this included the 'Big Four' Japanese bike builders. However, both the automobile and motorcycle industry's love affair with the turbo soon began to wain, this form of performance enhancement being virtually abandoned in favour of multi-valves and electronic ignition. Today, turbos are mostly to be found in a select band of up-market cars and heavy goods vehicles.

Twenty years ago it was the rotary engine which was the centre of attraction – an engine which owed its existence in no small measure to that other method of forced induction, the supercharger.

The man behind the rotary concept was the German engineer Dr Felix Wankel. During World War 2 his rotary valves were used in the Daimler-Benz aero-engines which powered aircraft such as the legendary Messerschmitt Bf 109 fighter employed by the Luftwaffe. After the conflict, Wankel became involved with the NSU company (see German Motorcycles, Osprey Publishing), building superchargers for the 50 cc streamliners which set a number of world speed records in the early 1950s.

Wankel convinced NSU management that the 'blower', which had helped to obtain an extra 800 per cent power from what had begun life as a Quickly moped engine, could be developed into a power unit 'of the future'.

That was early 1954; some three years later, in February 1957, the first Wankel engine – small enough to be held in one hand – was running. The Germans, never ones to overlook a potential engineering coup, soon had the world patent rights registered for the rotary concept, and wasted no time in selling manufacturing licences for the development and manufacture of Wankel-type engines in other countries.

Much of the initial income which NSU received went into designing and building the Ro80 automobile. This futuristic device combined a silky smooth 115 bhp twin rotor power unit, front-wheel drive, an electric clutch operated by a knob on the gear lever and an aerodynamically streamlined body shell that would not have looked out of place two decades later.

Unfortunately for NSU, the Ro80 was also hampered by a variety of problems. Heavy fuel consumption and rapid wear of the rotor tips were the two most cited causes of complaint. NSU spent millions of Deutchmarks in a vain attempt to rescue the project, but it was doomed. Bankruptcy followed; in 1969 NSU became part of the Volkswagen empire ...

Most of the Japanese auto industry (bikes and cars) had purchased licences. This included Mazda and Suzuki, the only companies brave enough or, in Suzuki's case, foolhardy enough, (and some would say the jury might

With a most unconventional engine, in an otherwise rather ordinary chassis, Suzuki's RE5 Wankel was a curious concoction of some of the finer elements of motorcycling, as well as some of the less satisfactory ones

still be out on Mazda) to put a rotary powerplant into production.

In contrast to Mazda, who down through the years have sold thousands of RX series cars, Suzuki's attempt was to prove one of the most expensive sales flops in motorcycle history.

The West German Cologne Show held in September 1974 saw three Wankel-engined motorcycles make their debuts. From the host nation came the DKW/Hercules W2000, powered by the fan-cooled Fitchel & Sachs 294 cc chamber engine. From Holland came the massive Van Veen OCR1000, a twin-rotor water-cooled giant and Suzuki's RE5.

Both the German and Dutch machines could boast a measure of pedigree: the Van Veen employing an automobile engine, while the DKW/Hercules employed a motor tested in snowmobiles over a relatively long period. Against this the Suzuki contribution was brand new and untried. However, the company was confident enough for six machines to be ridden non-stop from Los Angeles to Phoenix and back. Some two-dozen American journalists were unable to break any of the bikes, being flown across the country in relays to ensure the RE5's were kept going non-stop.

Sales to the public began in early 1975. At first glance, it looked big — and so it proved. The RE5 was a huge bike, tipping the scales at 240 kg (540 lb) and also suffering from an over long 1540 mm (60 inch) wheelbase.

The reason for this was quite simple, the bike was designed around the

The RE5 was heavy, thirsty and not particularly quick, but offered incredible smoothness, comfortable cruising and a whole host of detail requirements. It also lost Suzuki a fortune

engine. Unfortunately this had disadvantages, notably a heavy feel at low speeds thanks to a high centre of gravity and that over-long wheelbase.

The 497 cc engine was silky smooth but its 62 bhp at 6500 rpm, meant that performance wasn't outstanding when the high weight was taken into consideration. Motorcycle World achieved 13.7 seconds and 96 mph for the standing quarter mile in their August 1975 test. The same magazine said 'when pushed hard the RE5 delivers gas mileage that is downright depressing'. But in some ways this was unfair because testers of the day were judging the RE5's fuel consumption against other 500 cc bikes, when it should have been compared with at least a 750. Even so there is no doubt that high fuel consumption, high weight, high exhaust heat and the fact that motorcyclists on the whole are a fairly conservative lot, all transpired against the bike when it came to sales.

In Britain no amount of expensive publicity was able to offset the effects of the RE5's shortcomings and it soon became evident that the unconventional design was not going to sell. When it was finally withdrawn in 1977, fewer that 400 examples had passed out of the importers hands ... sadly for the RE5 and Suzuki this was a picture mirrored in every market in which the rotary engined machine was offered.

Custom Cruisers

Suzuki's first custom models, like their trail bikes, were originally little more than lightly converted roadsters, bikes such as the GS550/750L fours and the GS/GSX 250/400 twins of the late 1970s and early 1980s. They usually sported longer front forks, smaller rear wheels, peanut tanks and stepped seats. Cosmetic excercises which resulted in truly awful creations and evil handling.

By the mid 1980s Suzuki had finally realised that if it was ever to be successful in the 'Custom Cruiser' sector it had to offer specialised bikes.

The first of these were the LS650 Savage single (also offered in 400 guise for some markets), and VS750 Intruder V-twin.

Though outwardly similar to the DR600 engine, about the only thing the Savage mill had in common with its trailster brother was the oil filter. The 652 cc (94 × 94 mm) air-cooled sohc engine was in fact more closely related to the monster VS1400 V-twin which was launched shortly afterwards; the top end being the same except for a four-valve head and larger 40 mm Mikuni carb.

It might have said 'Savage' on the side panels but the 650 custom, as Motor Cycle News commented in a 1987 road test 'is about as brutal as a toothless dachshund'.

From a stationary viewpoint it has all the right credentials: clean, sculpted lines, pullback bars, spoked wheels, and acres of sparkling chrome

Left
Suzuki's VS750 Intruder was the first of the Japanese custom bikes to almost out-pose Harley-Davidson in their own back yard with its 45 degree V-twin engine and laid back styling

Right
650 Savage uses a 652 cc air-cooled single-cylinder sohc engine with four-valve head, but fails to deliver any serious performance

Odd-ball of the existing Suzuki custom family, the budget-priced GN250; low-tech and mediocre, it's more suitable for commuting than posing

and glossy metallic paintwork. Only problem is it also has several bad points as well. These include inadequate suspension, puny brakes and a rock-hard saddle.

Maximum speed is around 90 mph (with the rider tucked as low as he can get with those handle bars). The specification is completed by an electric starter, engine balancer shaft, belt final drive and a four-speed gearbox.

Finally, the lasting impression anyone will have of the Savage is its oh-so-low seat height – a mere 660 mm (26 in)!

If the Savage won't win any prizes for 'best in class', then the VS750 (uprated to 800 from late 1991) certainly will.

The Suzuki VS750 Intruder was the first of the Japanese customs to almost out-pose Harley in their own backyard with its 45 degree V-twin and clean styling.

The throbbing heart of the Intruder lies in its watercooled, single overhead cam, 4-valve engine. The following may be regarded as proof that Suzuki realised what was needed to take on Harley-Davidson: at first glance the engine appears to be air-cooled thanks to the deep finning on the cylinder barrels and heads, and the fact that the slimline radiator blends so well with the lines of the frame.

The 747 cc (80 × 74.4 mm) engine features a 45 degree phased dual-pin crankshaft design which not only manages to keep the engine's primary vibration to a minimum but also provides a balanced, gutsy, V-twin sound. Its silent-operation cam chains and self adjusting tensioner's reduce maintenance to a bare minimum, while transistorised electronic ignition and shaft final drive provide yet more advantages. Other details include an hydraulically operated clutch, 5-speed gearbox, 55 bhp (60 on 800), electric starter, 60-spoke wire wheels with polished aluminium rims, 12 litre fuel tank, 685 mm (27 in) seat height, dry weight of 199 kg (438 lb) plus a choice of high or flat handlebars.

Except for an increase in capacity to 805 cc (83 × 74.4 mm) and larger 36 mm carbs (formerly 34 mm) the current VS800 remains largely the same as the original.

The VS1400 is similar, except for size, to the VS750/800 series and made its bow at the end of 1986. But there are important differences. The massive 1360 cc (94 × 98 mm) engine has three rather than four valve cylinder heads, and no watercooling; instead there is a combination of aircooling of the front cylinder with Suzuki's race proven oil-cooling for the rear port. Hydraulic valve adjusters and automatic cam chain tensioners aid maintenance.

With all those cubes 71 horses may seem somewhat of a let down, but in practice low-end grunt more than makes up for this. It is debatable if even a four-speed gearbox − let alone the five ratios fitted − are really necessary.

Like all custom bikes handling and braking are not its strongest points but against this it manages a pretty good Harley impersonation ... and in the custom stakes that is what is all important.

This leaves the odd-ball of the existing Suzuki custom family, the GN250. It is a bike which belongs to the past, when Japanese custom bikes were tarted up roadsters. Motor Cycle News called it 'Custom Cecil'. The best thing that can be said about the GN250 is that it makes a good commuter machine. Custom cruiser? Forget it!

GSXR Series

The GSXR series of high performance sportsters was born out of Suzuki's participation in the world of endurance racing at events such as the Bol d'Or 24 hour classic during the early 1980s. The range eventually incorporated models of 250, 400, 600, 750 and 1100 cc, but only the two larger models are featured here -

The GSXR was launched at the Cologne Show in September 1984. The concept of the original GSXR750 was very simply to create the closest possible replica of the 1983 World Endurance Championship winning machine, but at a realistic price. Indeed, to make available to the public a machine which could, if required, convert their dreams of racing into reality, simply by fitting an optional factory-built performance kit.

The specification underlined the racing background of the GSXR; from the all new oil-cooled 100 bhp engine with its six-speed close ratio gearbox, through to the aluminium tube frame, and even the new Bridgestone endurance tyres. The styling was also pure endurance racer, warts and all. This was the real thing!

Right from the start the GSXR was built as a racer first, roadster second. The man who stamped his identity on the project, Yasunobo Fujii, saw to this. His vision was of a bike which could not only achieve a genuine 150 mph in standard road-going form, but also become an out-and-out racer with the fitment of a factory race kit developed in conjunction with 'Pops' Yoshimura.

The kit comprised a mouth watering assortment of performance goodies including a gas flowed head, high compression pistons, titanium valve springs, polished carb bellmouths, a dry clutch conversion, polished con-rods, high lift cams and a four-into-one race exhaust system.

Added to the standard engine, the 1985 model GSXR ended up with an extra 30 bhp. The result was a 170 mph hard charger which the likes of Honda's VF and Yamaha's FZ simply couldn't live with at the time.

Yasunobo Fujii was Suzuki's head of European engine development and as such he was one of the team that had led Suzuki into the four-stroke market with the launch of the highly acclaimed GS750.

Eight years on, the headline grabbing feature of the new 750 engine was its use of oil cooling; this was new in terms of motorcycle design, but had been used to help control the temperature of aero-engines since at least the early 1920s.

Suzuki first successfully experimented with the idea to cure overheating problems encountered during the development of the XN85 Turbo. Oil jets directed at the underside of the piston crowns prevented the turbocharged

GSXR750 represented state-of-the-art technology with its race-developed engine and alloy frame

engine from dropping a molten cocktail into the crankcase. The same idea was also employed on the earlier GSX750 EF motor to prevent the pistons in the two central cylinders from running hotter than their outside brothers.

While it would be accurate to describe the GSX motor as air cooled with oil assistance, the emphasis was switched the other way around in the 'R' engine.

Oil cooling came into its own in the new system which employed a second oil pump to concentrate on flushing away engine heat.

A sister unit mounted on the same shaft and sharing a common housing handled the lubrication. Each picked up from the sump and pushed oil through the engine at a rate of 20/22 litres per minute. The flow rate was in fact very similar to that of a conventionally lubricated engine, but the doubling up of pumps meant a doubling up of the

Above

The design of the GSXR750 was heavily influenced by Suzuki's participation in endurance racing events such as the French Bol D'Or with a series of prototype machines

Right

One of the factory entered GSXR750s in the 1985 Bol D'Or

circulation. The 5.5 litres of oil carried in the system was passed through a large cooler mounted on the front frame tubes.

Heat dissipation was further aided by the use of a magnesium cylinder head cover and a mass of shallow cylinder fins which gave the motor a distinctive appearance. All very clever; but why had Suzuki turned to oil cooling in favour of a water jacket?

The answer, claimed Suzuki, was to be found with a measuring tape and a pair of scales. The GSXR was both more compact and a lot lighter than would otherwise have been possible. A prototype using watercooling was found to be appreciably wider and some 5 kg (11 lb) heavier than the oil-cooled engine unit, which weighed in at just 75 kg (165 lb).

The search for power had led Fujii and his team down a more conventional path, with improvements to breathing and combustion coupled with moves to reduce frictional losses in the power chain.

The 749 cc (70 × 48.7 mm) 16-valve double overhead camshaft four-

More 1985 Bol D'Or action: GSXR (44) leads in this private battle

cylinder engine ran a compression ratio of 10.6:1 in standard form, while a bank of four 29 mm Mikuni flat valve carbs drew air in a direct path from an enormous eight litre filter box.

Inlet and exhaust valves were larger than those used in the EF engine, while the actual valve items were shaved down from
5.5 mm to 5 mm.

Higher lift cams and more radical valve timing was exploited, along with the introduction of a new, more efficient version of Suzuki's Twin Swirl Combustion Chamber (TSCC) design and a four-into-one exhaust system.

Close attention to reducing the weight of the piston end con-rod assembly allowed the crankshaft journals to be reduced to 34mm – 2mm less than on the 750EF. Crankshaft and con-rods weighed some 2 kg (4.2 lb) less than their counterparts in the earlier 750 engine.

The cam chain took its drive from the centre of the crank and was

Mick Grant, winner of the 1985 British Superstock Championship on a GSXR750

helped to keep on path by the introduction of an idler sprocket which also cut noise at high rpm.

New materials were used for the crankshaft bearings to reduce frictional losses in the transfer of power through the hydraulic clutch and six-speed gearbox.

Although Suzuki claimed otherwise, it was soon evident that the GSXR750 was saddled with a power delivery that was all-or-nothing, with little on offer in the sub-7000 rpm zone and whirlwind-type performance above it – hardly surprising when the engine's maximum torque of 49.2 ft/lb was produced at 10,000 rpm. All this made the GSXR something of a pig to ride in urban conditions. But outside city limits, or even better on a motorway or race circuit, the oil-cooled Suzuki came into its own.

The chassis was another area of conflict. Fashioned entirely from case and extruded aluminium, the box section, full cradle frame also incorporated a light alloy swinging arm and improved 'full floater' monoshock rear suspension.

Interestingly, the bike featured an 18 inch front wheel despite the factory having previously convinced everybody and his dog to adopt 16 inches in the interests of ultra-quick steering.

Somewhere, somehow, set against the obvious state-of-the-art technology which had gone into it, the GSXR was not without its handling faults ... riders on both road and track complained of high speed weaving, particularly on less-than-perfect surfaces.

But even if it did have a flick-switch motor and less than perfect high-speed manners, the new GSXR750 became by far the best selling machine of its class in 1985, a trend it continued into 1986, thereby helping Suzuki to establish themselves as a force to be reckoned with in Superstock and Superbike racing.

A year after the launch of the GSXR750 came a larger version, the 1100. The actual capacity was 1052 cc (76 × 58 mm) and, although largely unchanged from the original GSXR, the newcomer was a much more practical street bike.

Whereas the 750 provided power which was incredibly peaky, the 100's bigger cubes and reduced compression ratio (10:1) produced a truly flat torque curve throughout the entire range. This added up to respectable grunt from as little as 3000 rpm in top, massive mid-range at 5000 and awe-inspiring response by the time the 1100 reached the 750's lift-off point of 7000 rpm.

This only left a small question mark. That dreaded high-speed weave. And weave they did as anyone will testify who watched early GSXRs in full-blooded racing action at places like the Isle of Man TT circuit or Scarborough's Oliver's Mount course.

The 1986 GSXR1100 could reach 165 mph flat out; the standing quarter

Above

GSXR750L engine and exhaust system

Left

Suzuki made amends with the M model for 1991. This reverted to the original bore and stroke dimensions, had a modified frame and new upside-down front forks

mile was cracked in 10.8 seconds – a terminal speed of 126 mph.

Other changes from the 750 were 41 mm front forks (40 mm) 34 mm carbs 125 bhp and maximum torque of 74.5 ft/lb at 8500 rpm. The twin front brake discs were increased in diameter from 300 to 310 mm. There was also a very slightly longer wheelbase, a stronger frame and wider tyres to compliment the extra power and higher maximum speed.

In Britain both the GSXR750 and 1100 continued to largely dominate their respective racing classes in Superstock and F1, with Mick Grant finishing a glittering career in 1985 by winning the national Superstock championship title. Grant was followed by the likes of Phil Mellor, Roger Marshall and James Whitham.

Both GSXR models remained virtually unchanged, except for colours and graphics until the 1988 model year.

The 1988 GSXR750K came with larger valves and shorter stroke to produce a higher revving engine – 748 cc (73 × 44.7 mm). The double cradle MR-ALBOX frame was strengthened with a larger aluminium box

pipe. The motor however remained at the original 55 degree angle.

New Slingshot carbs, DAIS air intake system, a new twin swirl combustion chamber and Nissan four-pot brake calipers front and rear and floating double front disc brakes were all features of the modified bike, as were revised steering and suspension geometry, lower centre of gravity and the adoption of 17 inch wheels and Michelin radial tyres.

Suzuki claimed the chassis to be 'almost identical to the factory's Formula One and World Championship winning Endurance racers' whilst power was now up to 112 bhp at 11,000 rpm with a 13,000 rpm red line. The front fork stanchion diameter had been increased to 43 mm and the forks now featured 12 rebound and compression damping settings, a shorter swinging arm and 20, instead of 17 mm wheel spindles.

Suzuki engineers had concentrated on five points in the engine:
Increasing fuel charging efficiency.

1992 GSXR1100N. Unlike its smaller brother which received water-cooling, the 1100 retained the air/oil cooling of the earlier GSXR range

French Suzuki works team for the 1992
World Endurance racing series

Decreasing mechanical loss.
Increasing valve rev limits.
Increasing cooling efficiency.
Increasing engine response.

Air intakes at the front of the fairing directed cool air past the hot engine and a thinner, 20 per cent less resistant air filter into a more direct inlet port.

New flat slides in the 36 mm Slingshot carbs, much lighter and thinner, had a half round indexing ridge for an increased smoother flow.

A 2.5 mm increase for the inlet valves and a 1 mm increase on the exhaust side plus longer valve duration ensured more air reached the smoother Twin Swirl Combustion Chamber for increased power.

Mechanical loss and throttle response work together. The less the loss, the better the response. A 4 mm shorter stroke, lighter pistons, thinner

piston rings and a ten per cent increase in crank web diameter all worked in both departments to produce instant response and improved top end power. But at the same time, better materials, larger crankpins and journals ensured that reliability wasn't sacrificed. A 13,000 rpm red line demanded precise valve timing; this was achieved by using lighter rocker arms and larger diameter valve springs.

The 1988 bike employed larger diameter oil pipes which increased flow by 20 per cent while a bigger oil cooler and improved cylinder head flow increased cooling capacity by no less than 48 per cent. A cable operated clutch replaced the original hydraulically operated type.

Again the larger 1100 was a year behind in the development stakes, the 'new' model not being offered until the 1989 season, becoming the K model. This saw the capacity increased to 1127 cc (78 × 59 mm) while maximum power rose to 136 bhp at 9500 rpm. Many of the changes outlined above for the GSXR750 were incorporated into the larger bike. The chassis was new and increased rigidity by 25 per cent.

For all these changes the 'K' series GSXR's didn't really meet expectations and for 1990 were replaced by the L models ... and it was the smaller bike which benefited most from the modifications. Motor Cycle News summed it up thus: 'Yes, less is definitely more with the Suzuki GSXR750L. The gap between the 750 and the 1100 we tested last week has narrowed with the new 750L. So much so that in some areas, the 750 actually nudges ahead'.

The list of changes was a long one. as well as upside down front forks, there was more gusseting on the frame (using much from the 1100 type), an improved four-into-one exhaust, a steering damper, radial flow oil cooler, even larger (38 mm) carbs and a remote reservoir rear shock.

But the major change was to the engine. Suzuki abandoned the short stroke design of the 'K' and instead adopted the GSXR750RR (Racer) dimensions ... which just happened to be those of the original model launched back in the autumn of 1984.

The K model short stroke motor ran big ports and valves that didn't flow the fuel/air mixture well in the mid-range, and though it ran up to a stratospheric 13,000 revs, those same ports and valves couldn't give the charge the velocity, therefore density, it needed to produce more power at the top end.

So it had been a case of returning to the drawing board; the racer RR motor was basically the old design with refinements which enabled it to continue to achieve 13,000 rpm and make power. These included smaller

A 1992 GSXR750 works endurance racer with the fairing removed to show engine and chassis details

ports and valves, a 30 per cent more efficient oil cooling system, and stronger con-rods. Suzuki had learned a lot about these latter components after factory machines had suffered a succession of broken connecting rods in the endurance racing arena. Compared with the K, the GSXR750L produced far more usable power, which was particularly important for the street, if not the track.

A Motor Cycle News test commented that, 'There are no vibes, the clutch is light, the gearbox slick as a Swiss watch and the throttle so easy you find yourself holding it open inadvertently when you're braking'.

However, the riding position of the 750L was severe in the extreme. Even with wider, further back 'bars, most of the weight fell on the riders wrists. Quite simply you stretched over the tank and your legs were tucked up under your elbows if you were over 5 ft 6 in tall.

The frame had been stiffened all round with longer, thicker swinging arms, longer wheelbase, a bit more rake and trail and a massive 5.5 inch rear rim – the same size as the 1100, with a $170/60 \times 17$ inch radial tyre.

The end result was a bike which at last steered solidly with a nice, neutral feel, if a little heavier than previous GSXR models.

Suzuki had also tucked the four-into-one exhaust (now stainless steel) out of the way, along with the side stand.

More ground clearance meant that the GSXR no longer needed rock hard suspension settings to get round bends. The new upside down forks might have appeared a sales gimmick, but in practice they worked brilliantly, giving feedback to the rider whether he was braking, cranking hard into a corner, or just riding over a poor road surface.

As in the past there was a wide range of suspension settings, at both front and rear, to provide the optimum – if one had the time and inclination.

All in all, the 1990 750L was a great machine – and certainly the best so far in the class by Suzuki. However, many observers concluded that the 1100 was the best bet – it was less cramped, had more engine torque providing a far less frenetic ride; it also had a grab rail for the pillion passenger to hang on to, something the 750 didn't. Compared with the smaller bike, the 1100 featured a four-into-two exhaust system.

In 1991 the GSXR's were coded M, but with little change with the exception of the narrow fairing and bodywork, new dual-halogen headlight assembly and a die-cast aluminium instrument panel; there was also a wider seat on the 750 for improved comfort.

At the end of 1991, after seven years of GSXR production and much laboratory research and feedback from the endurance racing brigade, the Hamamatsu design team came to the realisation that it was finally necessary to make the switch to water cooling.

Since its introduction the combination of air and oil cooling had served

Upside-down front forks are a feature of the current GSXR750/1100 models, providing great torsional strength

the GSXRs well. It would be true to say that the bikes never suffered in terms of performance against the competition, having managed to hold their own in the relentless round of annual changes against other marques. The system even had some major advantages, such as greater simplicity in construction and a saving in weight. It was appropriate for the time. And yet when its time was up Suzuki had the good sense to realise it.

In designing the 1992 model GSXR750 (the WN), the factory at long last bowed to the inevitable and brought in water cooling (the 1100 remained air/oil cooled for another year).

And it was not just water cooling; the GSXR had been redesigned in almost every detail, without losing the structural proportions or characteristics of its predecessors.

Even if it appeared very similar to the 'M' model the chassis had, in fact, been completely redesigned in all-new 60 × 45 mm tubing, forming a continuous double cradle of forged and extruded sections with a detachable offside rail and bolt-on rear sub-frame. The 43 mm upside down forks remained unchanged, as did the link-type suspension, with its monoshock remote reservoir 'cushion' unit. There was however a new swinging arm including the use of different designs for the off and near sides. The near side was an aluminium box-section pipe measuring 71.6 × 38.6 mm, while the pressed-aluminium rigid one-piece off-side swinging arm was specially shaped to allow the exhaust pipe on that side to be tucked in closer to the centre of the bike. Combined with the slimmer bodywork, this feature allowed a deep banking angle of 54 degrees.

Revised weight distribution, with the engine being mounted lower down in the chassis and inclined forward at approximately 18 degrees, together with the changes in wheelbase length, trail and rake, resulted in a substantial change in the GSXR's handling. What didn't change was the wheels, tyres or brakes using the standard four piston 310 mm disc up front and single opposed piston caliper at the rear.

The same applied to the engine's bore and stroke dimensions, though virtually everything else was altered in some way during the change over to a water-cooled unit. The cooling system itself was served by a large, concave aluminium radiator, used in conjunction with the old SACS oil-jet system, in which nozzles spray the inside walls of the pistons to assist cooling. The main water cooling circuit also removes heat from the lubricant through a mini-radiator which goes round the oil filter base. Thanks to a total rethink of the internal engine components (lighter pistons and con-rods and a reduction in the crankshaft length, even though the journals are bigger) the watercooled engine is reduced almost 60 mm over the span of its predecessor to a very narrow 433 mm.

Another major difference from the air/oil cooled 750 engine is the camshaft arrangement which now opens and closes the valves by way of

bucket and shims operated by a pair of rockers, while the valve angle was narrowed from 40 to 32 degrees.

Apart from the much higher 11.8:1 compression ratio, the superior power output of the watercooled GSXR750 is derived from not only the weight saving of the pistons and con-rods, but from revised inlet ports and a new 6.8 litre airbox. Additionally, both the clutch and gearbox, with totally new ratios for the latter, have also been strengthened.

The sum of the revisions was to produce a more superior bike, not simply from a performance stance, but perhaps even more importantly, the 1992 GSXR750WN offered its rider much greater comfort and space to move, a result of raising the 'bars and at the same time lowering the saddle height (now only 780 mm). Although the 1993 (WP) 750 model was to remain largely unchanged, Suzuki updated the 1100 one year later.

So the 1993 GSXR1100WP is a significant improvement from the model it replaced, with not only water cooling, but also new six-pot calipers up front and a reduction in capacity from 1127 to 1074 cc; even the latter doesn't prevent a ten-horse increase in power output to 153 bhp in unrestricted form. Stunningly fast, and together with Kawasaki's ZZ-R1000 perhaps the last of the real superbikes before the legislators get their way.

Above
Jamie Whitham's 1992 factory GSXR750 racer with which he contested the British Championship series

Right
Endurance racing still plays a vital role in GSXR development, technology gained through track experiences being incorporated in future series production models

Off-road Competition

Of all Suzuki's off-road achievements one feat stands out: a decade of complete control of the 125 cc world motocross championship series from 1975 through to 1984.

The first ever 125 cc world motocross championship was held in 1975. Suzuki promptly took the title thanks to the combination of the most competitive bike and the talented Belgian rider, Gaston Rahier. Rahier retained the title for Suzuki in 1976 and 1977 but defected to rivals Yamaha for the following year.

Suzuki responded by putting Japanese national champion Akira Watanabe into the title fray and he repaid them magnificently by beating Rahier into second place for the 1978 world crown. Watanabe remains the only Japanese rider ever to win a world motocross championship.

For 1979 Suzuki took former 250 cc World Champion Harry Everts into the squad and the Belgian totally dominated the class for the next three years before handing the title over to fellow countryman Eric Geboers in 1982. Eric held onto the crown in 1983 and in a topsy turvy 1984 season Italian ace Michele Rinaldi grabbed Suzuki's 10th consecutive title at the final grand prix!

At that time Suzuki's success in the 250 and 500 cc world championship made them the most successful marque in the history of motocross with a total of 21 individual world titles in only fifteen years. They were also the only factory to have won a world crown in each capacity class.

But of course all good things have to come to an end and Suzuki's grip on the 125 cc world championship finally slipped in 1985 with the title honours passing to Cagiva of Italy.

It was not until five years later, in 1990, that the Hamamatsu company was to win another motocross world title, and then they did it in style, taking both the 125 and 250 cc classes.

The success of the works bikes out on the GP circuit has been matched by a ceaseless drive by Suzuki to cash in on this glory by offering paying customers 'over-the-counter' production motocross racers.

In the old days this stretched from 50 through to 500 cc, but currently Suzuki only offer three production off-road racers: the RM80, RM125 and RM250. All feature Suzuki's race proven Automatic Exhaust Timing Control (AETC). This system varies the height of the exhaust port to match engine rpm. A short exhaust port increases low and mid-range power, and a tall one increases peak rpm and high engine speed performance. The simple and effective AETC sliding valve blocks the top of the exhaust port at lower rpm, and recesses out of the way at higher rpm.

Late 1970s air-cooled RM series stormed the dirt bike racing scene

Technical details of the 1993 RM series are as follows:

RM80, 82 cc (47.5 × 46.8 mm), also 79 cc (46.5 × 46.8 mm) RM80X 27 bhp (26.5 bhp RM80X), Mikuni TM28SS carb, fuel/oil premixture 20:1, 6 speed gearbox, fuel tank capacity 4.5 litres (1.2 gal), dry weight 64 kg (141 lb).

RM125, 124.8 cc (54 × 54.5 mm), 39.5 bhp, Mikuni NTM 35SS carb, fuel/oil premixture 32:1, 6 speed gearbox, fuel tank capacity 7.8 litres (2.1 gal), dry weight 88 kg (194 lb).

RM250, 249 cc (67 × 70.8 mm), 54 bhp, Mikuni NTM38SS carb, fuel/oil premixture 32:1, 5 speed gearbox, fuel tank capacity 8.5 litres (2.2 gal), dry weight 98 kg (216 lb).

All feature single-cylinder two-stroke engines with water cooling and crankcase reed valve induction. Ignition is performed by Suzuki's Pointless Electronic Ignition (PEI) system.

Besides its involvement in motocross racing, Suzuki have also built enduro (PE) and trials (RL) machines, however at the present time only motocross is covered, the other two being deleted both at factory and customer level.

Off-road Suzuki has grown to mean motocross; the now legendary 'Yellow-Magic Force' is likely to continue as a serious contender at both world and national level as the new century beckons.

The year 1981 saw the RM125 equipped with water-cooling; performance didn't drop off as on air-cooled models

Above

'Berm-burning' with a water-cooled RM125X

Left

Production RM480; air-cooled and Full Floater monoshock rear suspension

Right

Vivid 125 action at Frome, Somerset, 1983

Water-cooled fours

Although there are others, two water-cooled two-stroke Suzukis stand head and shoulders above the rest; the GT750 and the RGV250.

This is for very different reasons; the GT750 three-cylinder model was the company's first series production 'water bottle', whereas today's RGV250 V-twin is at the very cutting edge of technology.

The older model was Suzuki's 1969 bid for a share of the action grabbed by the original Honda CB750 in the newly created 'Superbike' stakes.

No lightweight, the three carb, 67 bhp GT750 was capable of 0-60 mph in 5 seconds and top speed was around 112 mph.

Its engine was a 738 cc (70 × 64 mm) across-the-frame piston port three, with full unit construction and a five speed gearbox. Early models had a massive double sided twin leading shoe drum brake; later examples sported dual discs. The GT750 appeared in a host of model codes ending in 'B' when it was finally phased out in 1977, on the introduction of the GS four-stroke series, but not before it had spawned the highly successful TR750 racer.

Suzuki returned to the water-cooled two-stroke theme in 1983 with the parallel twin RG250 (247 cc and 49 bhp), followed shortly afterwards by the single-cylinder RG125 (123 cc and 12 or 19 bhp depending upon market), plus the square four RG500 (498 cc and 95 bhp).

First news of the trend setting RGV250 V-twin reached Britain in February 1987; the RGC260 going on sale in Japan a few months later. The first version to be exported was the 'K' in the 1989 model year, followed in 1990 by the 'L', which was virtually identical.

Both the K and L model RG250s shared the same 90-degree 249 cc V-twin engine with dual Slingshot carbs, a new air induction system, a radial flow radiator and Automatic Exhaust Timing Control (AETC) to increase power output right across the rev range. The result was 57 bhp at 11,000 rpm and a top speed of around 130 mph.

The high quality DC-ALBOX frame was made from cast aluminium to reduce weight while still maintaining maximum strength. There was the link-type Full Floater rear suspension with large 41 mm diameter fork stanchions, while braking was taken care of by twin 290 mm front discs and a single 210 mm disc at the rear. This system incorporated four pistons for each front disc, reduced to two pistons at the rear. Wheels were of the racing three-spoke type, 17 inch at the front, 18 inch rear. Dry weight was 128 kg (282 lb). These early model RGVs suffered from less than satisfactory

Suzuki's first production water-cooled motorcycles, the GT750

Left

GT750 was affectionately known as the water bottle or water buffalo

Above

RG250 Gamma set new standards when it was launched in 1984

reliability; broken/seized power valves and piston/cylinder seizures proved particular weaknesses.

For 1991 Suzuki engineers virtually redesigned the RGV in all departments providing a much improved machine. Not only was the engine uprated including completely new power valves (the original ones had been based on the motocross type), but there was a cresent-shaped, aluminium-alloy swinging arm. This latter component allowed both expansion chamber exhausts to exit on the offside and provide a 58-degree banking angle.

Other developments over the earlier RGVs included a coaxially mounted gearshift lever, as well as a needle roller bearing on the shaft for greater strength and altered gear ratios to improve acceleration. The main frame was also substantially improved, and the sub-frame modified for easier maintenance.

The front suspension now employed upside-down forks with a marginally

steeper steering head angle. Both wheels were now 17 inch, with wider section rims. Brakes too had been uprated; front discs being enlarged to 300 mm. Power output at 59 bhp was not much different, but a reprogrammed engine management system provided more useable power characteristics. This showed not only in normal street use, but most notably on the race circuit, where the RGV250M became the bike to beat in 1991 Super Sport 400 class racing.

This success was continued into 1992 with the 'N' model and subsequently the 1993 'P'. Both were still able to hold their own with the very latest 400 four-stroke four-cylinder models such as the RC Honda, ZXR Kawasaki and FZR Yamaha. As an example of this the hotly contested British Super Teen championship, Scotsman Callum Ramsey secured the coveted title on a 1991 RGV250M, proof, if any was needed, of the tiny V-twins true potential.

Above
The RG250 Gamma was soon followed by the 498 cc RG500 Gamma

Right
Mark 2 Gamma looked like this; a 1988 Limited Edition in Pepsi colours shown

Left
Pulling wheelies was all too easy on the RG500 Gamma

Above
RG500 Gamma, with tank, seat and fairing lowers removed

Above

RGV250M introduced for 1991 featured gull-shaped alloy swinging arm, twin silencers exiting on offside and upside-down forks – plus numerous technical changes

Left

In layout and likeness the 1985 RG500 Gamma street engine was remarkably similar to the pukka race engine. The square four rotary disc induction motor produced a class-leading 95 bhp

Brightly painted RGV250M raced by Andy Murphy under the York Suzuki Centre, Cadwell Park 1992

Above

Four RGV250Ms awaiting the off at Cadwell Park Super Teen round, August 1992: Gary Walker (33), Brian Moffat (69), Sam Harrison (39) and Michael Pateman (12)

Left

Gary Walker's RGV250M Super Sport 400 class racer at Cadwell Park, August Bank Holiday Monday 1992

Above

An RGV250M in its natural environment – the race circuit

Right

James Hayden rode this RGV250M in 1991; it was then purchased by former racer Mick James (left) for his son Carl (right) to compete in the 1992 MCN Super Teen series

SUZUKI